Canyon

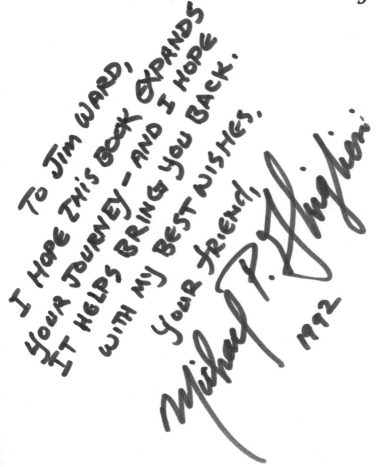

To Jim Ward,
I Hope This Book Expands
Your Journey — and I Hope
It Helps Bring You Back.
With My Best Wishes,
Your friend,
Michael P. Hughes
M 92

Canyon

Michael P. Ghiglieri

The University of Arizona Press
Tucson & London

The University of Arizona Press
Copyright © 1992
Michael P. Ghiglieri
All rights reserved

♾ This book is printed on acid-free, archival-quality paper.
Manufactured in the United States of America

97   96   95   94   93   92   6   5   4   3   2   1

Library of Congress Cataloging-in-Publication Data

Ghiglieri, Michael Patrick, 1946–
    Canyon / Michael P. Ghiglieri.
        p.   cm.
    Includes bibliographical references (p.) and index.
    ISBN 0-8165-1258-2 (cloth). — ISBN 0-8165-1286-8 (pbk.)
    1. Grand Canyon (Ariz.)—Description and travel.   2. Geology—
Arizona—Grand Canyon.   3. Rafting (Sports)—Colorado River
(Colo.-Mexico)   4. Ghiglieri, Michael Patrick, 1946–
—Journeys—Arizona—Grand Canyon.   I. Title.
F788.G45   1992
917.91'320453—dc20   91-29222
                                                                      CIP

British Cataloguing-in-Publication Data
A catalogue record for this book is available from the British Library.

To Cliff and Crystal

The Canyon still grips its travelers in mysticism. No one is cynical or matter-of-fact about the Grand Canyon. It is perhaps the world's greatest natural wonder. There is so much in the experience of being down in it that cannot be grasped. Its size, its age, its power, its beauty are so grand, so completely overwhelming, so beyond comprehension that the human brain is overpowered, numbed. Previous perspectives are reoriented, experiences negated, conclusions doubted. The Canyon becomes a mind-altering drug, the canyoneer an addict. He wonders if ordinary people can understand what he's experienced; he becomes "born again," sharing a bond only with others of the same persuasion. It becomes "his" or "her" canyon. Only the initiated can truly understand the power of this fanatic possessiveness.

Bill Beer

Contents

GRAND CANYON

The Canyon from Mile 0 at Lee's Ferry to Mile 100

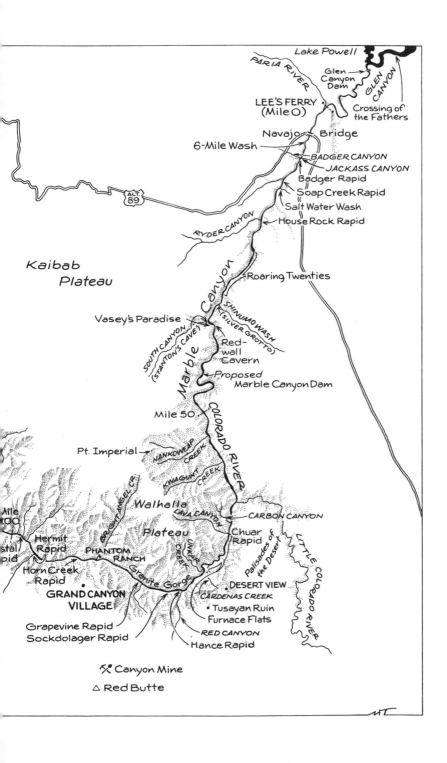

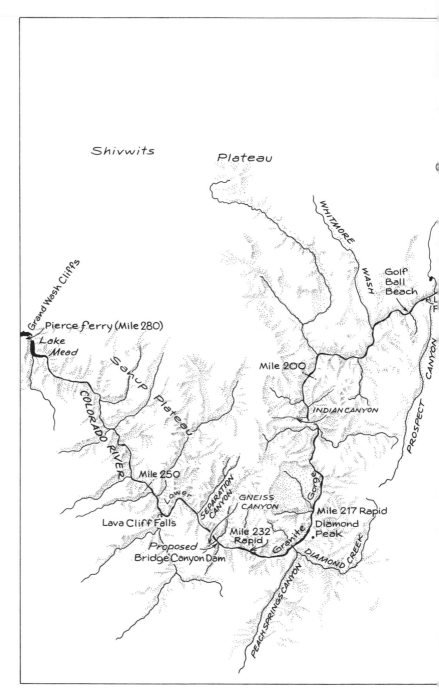

The Canyon from Mile 100 to Pierce Ferry at Mile 280

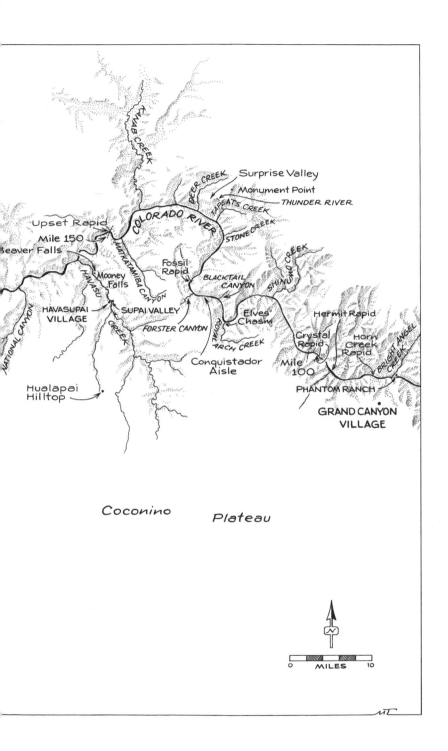

Surprise Valley

Monument Point

THUNDER RIVER

DEER CREEK

TAPEATS CREEK

COLORADO RIVER

STONE CREEK

KANAB CREEK

Upset Rapid

Mile 150

Beaver Falls

MATTKATAMIBA CANYON

Fossil Rapid

Mooney Falls

HAVASU

BLACKTAIL CANYON

SHINUMO CREEK

Hermit Rapid

HAVASUPAI VILLAGE

SUPAI VALLEY

Elves Chasm

Crystal Rapid

Horn Creek Rapid

CREEK

FORSTER CANYON

ROYAL ARCH CREEK

BRIGHT ANGEL CREEK

NATIONAL CANYON

Conquistador Aisle

Mile 100

PHANTOM RANCH

Hualapai Hilltop

GRAND CANYON VILLAGE

Coconino   Plateau

0   MILES   10

Preface

Running the Colorado River in Grand Canyon is to me the most impressive journey on our planet—an adventure that leaves no traveler unchanged. For seventeen years between ecological research projects, writing, and teaching I have worked as an international whitewater guide, mostly here in the Grand Canyon of the Colorado. During those years I have searched bookstores for *the* book on the Canyon. Perhaps because nearly all existing books were written by outsiders or casual visitors without the time, interest, and experience to do justice to this, the most impressive of our planet's Seven Natural Wonders, I never found it. Finally I realized that if it was to be, I would have to write it. So here it is: the book on Grand Canyon that, during my years as a professional guide, I always wished I could find.

This book is written as a journey—one that really happened—down the Canyon. Along the way are the human adventures and mysteries of the Canyon's exploration, battles for its conservation, portraits of the Native Americans who once dwelled—and still do dwell—in its depths, plus wildlife and a revealing look at the huge chunk of planetary history written in its rock. All of these are woven into the real adventures of twentieth-century guides who take passengers for hire down pounding white water roaring through the last major region in

North America to be explored. Welcome aboard. Fasten your life jackets.

I owe a debt of gratitude for encouragement and help in shaping this book to Stephen F. Cox and to Robert Asahina. For assisting in my research on facets of it I owe Marty Anderson, Jan Balsom, George Billingsley, Gary Bolton, Nancy Brian, Susan Cherry, Kim Crumbo, Dan Dagget, O'Connor Dale, Regan Dale, Helen Fairley, Kenton Grua, Ivo Lucchitta, Deborah Moore, Scott Thybony, Beth Trepper, Mike Walker, Dave Wegner, and Michael Yard. As a special debt, I owe Bob Hoffman thanks for his help in wrestling with my word processor and beating it two out of three. While I readily acknowledge my debt to them, none is responsible for the content of this book. Any shortcoming it may have is my responsibility alone.

Canyon

1
A Monster Is Born

Adrenaline Alley—forty miles of monstrous white water roaring a mile deep in the earth's crust. These forty miles of First Granite Gorge are the reason why some of us cannot stop coming back here. They are also the reason why some never return. At this point we had successfully navigated its first twenty miles. Unfortunately, what lay ahead was far worse.

Waves below the horizon of Crystal Rapid exploded like claymore mines. What worried me most was the hidden gauntlet of three killer holes into which the Colorado River funneled immediately ahead. The one good thing about these dangers being hidden from me—and from the other four people in my boat counting on me to get them through all of this safely—was that the holes ahead cannot hurt you until you can see them. The bad thing was that we were heading right toward them.

This sudden transformation of the Colorado from a moving pond to a liquid predator here at Crystal Rapid automatically triggered a release of adrenaline. I felt my heart pound faster and the back of my neck tingle. My perception of time warped and the seconds dragged. Crystal is the drug of choice for adrenaline junkies. Even so, it made little sense for me to be here now. No, I was still in good health; I could handle the oars as well as ever.

I was not here because money was a problem. Nor were things bad at home. In short, I did not have to be here leading this trip. Indeed, I knew I could lose everything important to me by continuing to court this monster of a river. The simple reality was that I could not stay away.

Abruptly, the entire huge mass of the Colorado River accelerated downward. My stomach rose as if descending in an elevator whose cable has snapped. I had no time to contemplate this. For two miles the river had been as smooth as glass, dammed to stillness and anticipation by the spew of boulders that made up Crystal Rapid. That had been the time to think. Now I was dropping over that dam into the last seconds of sanity in the transition zone of the rapid's tongue—the long, V-shaped slick funneling into the deepest part of the rapid. Now was the time to row, the time for inspired action. I aimed the stern of my boat downstream toward the right. With the demon of failure whispering in the back of my mind, I pulled on my oars as fast and hard as I could to escape. This used to be fun, but it was not fun now. Not since 1983.

Actually, until 1966 Crystal was more a large riffle than a rapid. But on December 5 and 6 a monster winter storm dumped fourteen inches of rain into the drainages of Crystal (Mile 98$^+$, that is, more than 98 river miles downstream from Lee's Ferry, which is at the head of Grand Canyon National Park) and Bright Angel (Mile 88) creeks within thirty-six hours. Fourteen inches was the normal precipitation for nearly two years here. This deluge created an erosional event that roared down Crystal Creek like a blitzkrieg of panzers into Poland.

The capacity of moving water to carry a load of ever-larger particles cubes as the speed of the current doubles. This means that really fast water can carry very large and heavy objects suspended in its flow. It can also bounce even bigger ones along its bed. Because so much of the surface area within Grand Canyon is innocent of soil, most of those fourteen inches of December rain soon penetrated to bare rock and then ran downhill in a slurry carrying more rock and mud than water. For a few hours the flash flood in Crystal Creek transported boulders the size of Volkswagens. It spewed its seemingly endless load of boulders into and across the Colorado with such energy that some of it tumbled against the opposite wall of Precambrian schist—rock that was ancient a billion years before the first worm wriggled through primordial ooze.

It is likely that no flood this big—at least 10,000 cubic feet per second (cfs)—had roared down Crystal Creek for millennia. Certainly nothing like it had happened in the past nine hundred years, because this flood scoured away Anasazi ruins that had stood preserved in Crystal Canyon (supposedly well above flash-flood level) since the heyday of the Pueblo II culture in Grand Canyon.

It had been one hell of a flood, one that redesigned and transformed Crystal Rapid into a whitewater challenge. But because the Colorado was interrupted and regulated upstream by Glen Canyon Dam, the river lacked the muscle to put its normal polish on this infant monster. The constrained flows of the Bureau of Reclamation's new controlled Colorado feebly plucked at the smaller boulders in the "dam" at Crystal but could not budge the larger ones. The result was a rapid that most boatmen ranked number two in Grand Canyon, just behind Lava Falls. Although at certain water levels other rapids are worse (Horn Creek and Hance rapids at medium-low flows, for instance), Crystal generally stacked up as number two. For the seventeen years after the new Crystal formed, none of us knew how lucky we were that Crystal Rapid was only number two. In fact, during the second half of those years, when I was rowing professionally, Crystal never seemed difficult to me. I had learned to row on rivers that offered a lot worse.

But in June 1983 all of that changed. The Bureau of Reclamation botched their storage planning at Glen Canyon Dam and trapped themselves with a full reservoir on June 2, before the Colorado River peaked for the season. Myopia in the bureau and greed among the brokers of hydroelectric power had produced a policy of storing additional water in Lake Powell through March, April, and May, thereby leaving only five vertical feet of storage remaining to contain the big snowmelt. This was despite the National Weather Bureau's warning that runoff would be above normal. Impending floods were so obvious in Utah that in February Salt Lake County made room in its reservoirs and warned citizens to prepare for high runoff. Even more inexplicable, the bureau gave up power revenues from Glen Canyon during the first five months of 1983 by releasing only medium flows through its turbines in order to retain this water.

For perspective, one must recall that in the early 1980s, during his first term as president, Ronald Reagan both slashed taxes and escalated defense spending. In an attempt to balance this, Reagan turned to the only other branch of government besides the Internal Revenue Service that generates significant revenue—the Department of the Interior. In 1983 the Interior Department, especially the Bureau of Reclamation, was still under pressure to maximize revenues. Does this explain why the bureau so parsimoniously saved early runoff? Glen Canyon Dam had been built to store water, not to control floods, and its hydropower did pay the bills—nearly a billion dollars' worth between 1964 and 1990. Maybe the bureau was gambling that the water they saved in spring would generate more valuable power during the summer.

When the river did peak in June at about 120,000 cfs, the bureau was forced to match outflow with inflow. This seemed to pose little problem. The dam had ostensibly been designed to dump more than twice that. The two spillways alone, each forty-one feet in diameter, were supposed to channel 276,000 cfs. But by 1983 few in the bureau remembered that when the plans for the dam had been hashed out in 1958 the architects and budget men had been faced with alternate plans for spillways. Their choice had narrowed to two designs: one costing $5 million more than the other (in 1959 dollars). You guessed it. The bureau decided to whittle Glen Canyon's price to $70 million and save the public $5 million by building the cheaper spillway tunnels, which, they suspected, would rarely be used because Lake Powell would rarely fill to Glen Canyon Dam's full height.

When, in late June 1983 Lake Powell finally did reach an elevation of 3,700 feet, the capacity of the 710-foot-high dam, the bureau opened the spillways and let her rip. But what shot out from the far end of the east spillway at 32,000 cfs made the engineers gulp in fear. The pretty spouts of white water abruptly turned orange with sandstone and shot out spinning slabs of concrete, the linings of the spillway tunnels. The trouble was that Glen Canyon Dam had been emplaced in bedrock that made good engineers avert their heads, as if seeing a drunken friend making an ass of himself. The soft Navajo and Kayenta sandstones not only were porous but also eroded like public support for a crooked politician. The engineers knew that once the linings within the tunnels were breached, the river's tumbling and cavitation within them would gouge immense holes in the bedrock, and depending on where those cavities were now forming in the 3,000-foot-long tunnels, this erosion could compromise the entire dam.

Worse, this would not have been the first time the bureau had lost a major dam. Seven years earlier, in June 1976, Teton Dam in Idaho had collapsed. Due to the usual reasoning connected with pork-barrel politics, it had been emplaced in obviously unsuitable bedrock despite an advance warning from U.S. Geological Survey geologist David Schleicher and two of the bureau's own geologists. Overriding additional opposition from environmentalists, the bureau spent $85 million to build the dam. It lasted seven months. When the 305-foot-high dam failed on June 5, the Teton River suddenly flowed at two million cfs, roughly equal to the Congo River. The flood rolled over the towns of Wilford, Sugar City, and Rexburg and almost took Idaho Falls as well. Amazingly, because the dam had failed during the day and was spotted in time to send warnings downstream, only eleven lives were lost. But damage to property totaled $2 billion, not including the stripping of topsoil from tens of thousands of fertile acres (Marc Reisner gives us an excellent account of this man-made disaster in *Cadillac Desert*). After the flood the town of Wilford ceased to exist—new maps no longer list it. And Teton Dam had been only a medium-sized dam.

The bureau's options at Glen Canyon Dam had radically dwindled. If they continued to operate the disintegrating spillways, they risked losing the dam. If, instead, they closed the spillways, Lake Powell would overflow and destroy much of the dam. So the bureau resorted to raising the effective top of the dam by installing four-by-eight-foot sheets of plywood to help hold back more than 186 miles of impounded river—the world's second longest reservoir after Egypt's Aswan Dam—containing a total storage of 27 million acre-feet, two years' flow of the Colorado.

Disbelieving, we in the O.A.R.S. crew drove to Page to see the plywood for ourselves. This flimsy three-quarter-inch barrier atop 10 million tons of gleaming white concrete poured into a monolithic arch was not reassuring; we would be rowing the river immediately below it for the next two weeks. But even though the plywood held because the horizontal pressure of the water at the surface was minor, the bureau replaced it with eight-foot-high steel barriers.

The engineers also fine-tuned the flows to the point where concrete ceased spinning Frisbeelike into the river below. They pinched the outflow from the east spillway to about 10,000 cfs, the west to 39,000 cfs. Meanwhile, about 48,500 cfs howled through the channels within the dam itself to feed the eight turbines of the power plant (33,100+ cfs) plus the river outlet works (15,000+ cfs). But the Colorado River was now roaring into Lake Powell from Cataract Canyon and from the San Juan River at 120,000 cfs. In a race against Mother Nature, the boys of the bureau watched Lake Powell inch its way up to the top of the barrier at 3,708.4 feet. Two weeks later one senior federal employee working there confided to me, "No one knows how close we came to losing that dam."

Had the dam failed, it might have taken out the next eight dams between it and the Gulf of California. The death toll would have been staggering.

By July 1 the bureau had won—barely—but the cost of victory had been high. The domino effect flooded eight dams downstream, including Hoover Dam, causing millions of dollars in property damage. Repairing Glen Canyon Dam's damaged spillways—now enlarged by the flood with new rooms at those "cheaper" elbows that looked like Carlsbad Caverns—and installing new air slots in the spillways to aerate the flows to reduce cavitation cost $35 million for the construction alone. The bureau called this "a triumph of the human spirit."

But environmental damage within Grand Canyon was extensive and irreparable. Surprisingly, just as in the follow-up to the disaster at Teton Dam—a dam that never should have been built—the bureau fired no one after the Glen Canyon Dam fiasco. Instead, Robert N. Broadbent, commissioner of the bureau, blamed their computer program for mismanagement.

Perhaps the scariest product of the bureau's sudden dumping of Lake Powell was the remodeling of Crystal Rapid. The bureau's reincarnation of the wild Colorado carved a deep new channel along the left side of the unstable "dam" created by the 1966 flash flood debris flow of Crystal Creek. It tripled the dangers in the rapid and made the "break-out" to avoid these dangers an order of magnitude harder. Motor rigs more than thirty feet long and carrying twenty people flipped end-over-end. So many motor rigs flipped or were stripped of passengers in Crystal Hole that, between June 24 and 25 alone, the helicopter rangers of the Park Service had to pick up ninety-one survivors–a dozen of them injured and one dead—and chopper them out of the Canyon. This was a record for the Canyon "Danger Rangers" of the National Park Service. Even the "Woman of the River," Georgie White, motoring the largest raft to float the

Colorado, lost every passenger and nearly all her gear—stripped from her "Queen Mary" into the maelstrom of Crystal Hole. When asked why, she explained: "They don't make passengers like they used to."

Kim Crumbo, a friend and principal ranger, described the mayhem to me as a nightmare. He was no stranger to hazard. As a Navy SEAL in Vietnam, Crumbo had been dropped into bizarre missions. Crumbo and I had also spent thirty-five days rowing down the Omo River in Ethiopia. One boat had flipped and had tossed its passengers into water swimming with crocodiles known to be man-eaters. As we rescued those swimmers, he had seemed blasé compared to his attitude now about Crystal's menace.

On July 1, 1983, we tied our boats to boulders upstream of Crystal to scout it. The flow was about 90,000 cfs. The twenty-foot wave recycling into Crystal Hole was so colossal that it seemed an illusion. Surprisingly, the rapid was not as difficult to run then as it would become as the spate ebbed. These days, at flows above 30,000 cfs, Crystal is by far the toughest on the river (although nothing like the white water that would have been created had Glen Canyon Dam failed). And in 1984 and 1985, the rapid became even harder.

Before rowing it this time, we spent an hour studying its every nuance. We focused on the tiny alley of safety along the right shore. That day, at 38,000 cfs, all the water entering Crystal was rushing to the left like a King Kong conveyor belt into mayhem: a chain of dangers, each link of which had earned a name immediately after its creation. The tongue dived into the New Wave, zigged left malignantly into the crushing Slate Creek Diagonal, then pounded roller-coaster-like into Crystal Hole. There was no question of maneuvering through these, anymore than a fly could maneuver down a flushing toilet. Failing to make the power cut out of the tongue to hug the right shore placed a boat at the mercy of each danger. One led to the next. The odds of flipping in any of them were high; more boats upset here at high water than in all the other rapids of the Grand Canyon combined. One company had flipped three out of four boats here at half this flow. Gigantic motorized pontoon boats had upset here in less water than was present that day, scattering passengers into liquid insanity. It still happens too frequently.

But Crystal's worst dangers follow this gauntlet. While flipping a boat brings the risks of hypothermia or being trapped beneath the boat and drowned, the worst risk is being slammed or crushed between the drifting boat and what lies downstream of Crystal Hole. The second half of Crystal is a slightly submerged island of boulders called the Bone Yard, the Dory Graveyard, the Rock Garden, and other similar appellations. It has broken or bruised several of the hundreds of swimmers tossed from boats that had flipped in the New Crystal. A few have died.

The only hope of avoiding this navigational menace is to row as hard as possible at the top to break out to the right of the sweeping current of the tongue as soon as possible—but without hitting the shore and bouncing out to be swallowed by the New Wave. This is simple but not easy. The only other option is to portage, which would take all day. That day this break was possible in only one tiny place, adjacent to a bulge of water atop a marker rock that was invisible until one had rowed too close to make any real adjustment in the boat's approach. This is what made the rapid so tough. It required full-strength rowing toward the right against an extremely fast current and diagonal waves funneling to the left—one must power the boat precisely toward an eddy the size of a throw rug behind that one tiny marker whose position one had to imagine until too close to do anything about it. It was not fun.

Scouting this rapid to identify the ephemeral little "landmarks" in the current that were critical to figuring out one's timing to start breaking out of the tongue was nearly as unnerving as rowing it—sometimes more so. During seventeen years as a professional guide I had seen boatmen come apart at the seams here. I had seen others go out and row heroically when they were so scared that they first had to sprint for the bushes seconds before rowing to evacuate their bowels in a race against a sphincter they could no longer control.

Perhaps the toughest scouting I had had to witness was during one of many trips here with fellow guide Michael Fabry, a friend of a dozen years and a hundred trips. He had flipped here at higher water on his previous trip, losing his boat at the top in the New Wave. On this, Fabry's next trip I had watched him for twenty minutes as he studied the rapid, made up his mind about this or that section, then rechecked it twice. He knew he was lucky to be alive, and he suspected that rowing Crystal again might be pushing it too far.

Crystal on Fabry's previous trip had become a nightmare even before he rowed away from the beach. The river was high, more than 30,000 cfs, too high to carry passengers with any margin of safety. As trip leader, he had followed the standard policy, explaining to the passengers that the crew would row the boats empty on the cheat run while they walked around the rapid. One of Fabry's passengers objected, hissing that he had paid thousands of dollars to bring his family on this trip and that he was going to get what he paid for. Fabry explained that walking around the rapid was not an arbitrary decision but the recommendation of the Park Service at this flow level because of the extreme danger. The man became abusive. As Fabry rowed away from the beach, his passenger cursed him with a spray of obscenities that would have blistered the paint in a men's locker room.

Fabry lost his concentration—and blew his entry. The first big diagonal wave grabbed his boat and shuttled it into the New Wave, where the giant hydraulic flipped it upside down as if it were crushing a fly. Suddenly the shore seemed the most desirable place in the universe but impossibly distant.

Behind him, Michael Boyle entered the rapid pulling hard, a surge of adrenaline driving his boat into a perfect entry. Boyle had seen Fabry's boat pirouette, then flip. Normal protocol called for Boyle to start his chase to rescue Fabry and then his boat, but immediately behind Boyle, Renee Goddard was rowing into the rapid. She was new, and Boyle knew that if Fabry had flipped, Renee likely would too. Boyle's judgment was colored by the soft spot he had had for her ever since the Waghi-Tua-Parari exploratory trip in Papua New Guinea three years earlier. Now, instead of chasing Fabry, Boyle pulled toward shore below the New Wave to watch how Renee fared.

Renee's boat appeared. Her entry was weak and not far enough right. Her boat vanished in the New Wave, then reappeared upside down. Spurred by the realization that he now had two people to rescue and two boats to chase—all scattered many seconds apart—Boyle pulled on his oars to intercept Renee before she was swept into the Rock Garden. This left Fabry facing a double nightmare: a flip as lead boat and no boat chasing him for rescue. (Recently a guide flipped here, and because the other three boats with him were committed to collecting passengers, he had drifted alone for eighteen miles, clinging to the floor of his upside-down boat. Hours later, after being rescued, his first words upon sitting down on terra firma were, "It sure is stable.")

To make short what seemed to Fabry an endless story, he got pounded. He plunged through a quarter-mile gauntlet of abuse in bitterly cold water (only 52° F). Hypothermia paralyzes muscles quickest in a rapid because the jumbling hydraulics constantly strip away the water adjacent to a swimmer's skin before it has a chance to become a warmer insulating layer. Bad as this was, Fabry's survival was compromised by yet another, unforeseen problem.

The crushing hydraulics and conflicting currents had torn his life jacket to shreds. He was spun like a window shade underwater, then agitated as if trapped in a giant Maytag in the Slate Creek Diagonal—and then again in Crystal Hole. When he finally surfaced en route to the Rock Garden, sections of ensolite flotation from his disintegrating life jacket bobbed up near him—a clue as to why he was floating so deep in the water. His boat was now too distant to reach. No way could he haul himself out of the pounding waves and onto its upside-down floor.

Fabry knew the submerged island was coming, but, already dazed, oxygen-starved, and mildly hypothermic, he could not swim clear of it. A decade earlier, living in a tepee along the Stanislaus River, Fabry had swum small rapids naked each day to perfect his technique. Perhaps this is all that saved him now. Two or three seconds after Fabry realized his life jacket was disintegrating, the Rock Garden pounded him unmercifully. Twice it drove him to the bottom of the river, where he slammed against the boulder-strewn bed. During his brief moments on the surface, he fought for air by breathing only on the downslope of each wave. And he fought to keep himself facing downstream, shallow and horizontal, but being so deep in the water due to lost flotation made it almost impossible. The roaring tumult of frenzied waves yanking him under did make it impossible. He hit bottom in yet another hole. His shin slammed into a boulder and he felt the bone crack.

Eventually he was swept beyond the foot of the long rapid, where he stroked weakly into an eddy and crawled out on the right shore. On the far side of the river, racing toward Tuna Rapid, was his capsized boat. On it his ammo box had sprung open, dumping into the river his camera and a $350 tip for the crew from passengers who left the trip upstream the previous day at Phantom Ranch. Meanwhile, Boyle rescued Renee before she was swept through an instant replay of Fabry's "swim." Bruce Helin, the fourth boatman on that trip, pivoted his own rig and chased Fabry's boat.

Now, on Fabry's next trip, I watched him scout. Instant replays of his swim flashed through his mind as he struggled to evaluate the rapid calmly. For Fabry, scouting now consisted of watching reruns from hell.

"Got your route picked?" I asked, wishing that I could make

it easier but knowing there was no way. All performance was strictly solo out there.

He shouted over the roar, "Just break in here," and gestured toward a little valley on the first major diagonal wave adjacent to a chair-sized boulder we used as a marker rock when we rowed close enough to see it. Here we were so close we could have spit on it. "And keep her straight but spin around if my nose gets pulled out there," he added, pointing toward the second major diagonal, a much bigger wave that, at higher water, stalls almost every oar boat that hits it and then diverts each into the New Wave.

It was a reasonable plan. It was the only plan. I deliberated over probing further to see if I could suggest anything to help, but I was afraid I might say something that would backfire, with his misconstruing it to mean that I thought his plan was off. His anxiety was too high. We had worked on rivers together long enough to become intimate and long enough for me to know that he knew how to row. But not long enough for me to overcome his fear of getting trashed again in Crystal. A hundred years could not do that.

"Good plan," I nodded. Then I asked, "Gonna wear your glasses?" and instantly regretted it. Fabry thought that glasses ruined his Marlboro Man image by making him look too studious, which was ironic because he held a teaching position in cardiac rehabilitation. His habit of running the river without his glasses and missing some of the right places and hitting a couple of the wrong ones had prompted our fellow guide, T.A., to nickname Fabry "Mister Magoo." Anyway, hitting a wrong one here . . .

"At least there are no crocs or hippos," I added in a lame attempt to cancel my remark about glasses by painting the rapid as such a cinch that we needed somehow to spice it up.

Fabry turned toward me for the first time and grinned. "That's what we really need—another hippo attack." He snorted in reminiscence, then said, "Yeah, I'm wearing glasses." He returned to gazing at the river, then added, "I'm ready. Let's get the ducks on the pond."

We did, and he had aced it. But that was then, and on medium water. This was now, and the river was screaming between the jagged cliffs at 38,000 cfs, and I was lead boat, making the critical entry on a very ugly slice of white water.

We dropped over the horizon. I scanned over my shoulder and finally spotted that little boulder. I concentrated on getting the most out of my last stroke. One more stroke would shatter my left oar against that marker, turning it into flying splinters of plastic and fiberglass. I needed both oars to row the 90 percent of the rapid remaining. Rowing lead boat meant, as it had for Fabry, that my four passengers and I had virtually no chance of being rescued if I blew it here.

The marker flashed past and instantly descended in importance from critical to zip. We plowed squarely into the first diagonal wave, slowed suddenly, then rammed into the second diagonal—a real boat stopper. It absorbed our momentum like a sponge, redirected us left, and funneled my Domar riverboat onto the right shoulder of the New Wave. This was not in my plan.

I yanked on my right oar and shoved on the left to pivot and spin the bow, with the combined weight of the three passengers up there, vital to climb straight up the New Wave and punch through it. I yelled "Forward!" to signal Gloria, Wilson, and Susan to scramble to the tip of the bow, where their weight would act to pull us over that wave. John, in the stern, gripped the rigging straps as I had coached him. Although the boat was now speeding so fast that it made the river blur to them, my rush of adrenaline had shunted me into a time warp that stretched out the split second it took those three to make that essential leap to the bow and for the raft to climb the shoulder of the New Wave. It seemed as long as a bank line on payday.

Hundreds of gallons of Colorado River water exploded over us. During this whiteout I felt my left oar crab in the current and shove deep into the base of the wave. Instantly I knew my eleven-foot oar had gone beyond the point of no return. This was not part of my plan either, and although I had seen this happen dozens of times, it had happened to me only once before. I felt a prick of irritation for having allowed it to happen again.

"Shit!" I heard myself hiss. Unless I let go of the oar very soon, it would either snap in two at the oarlock or catapult me out of the boat into the drink. Neither was desirable. If I released it, the oar would pop out of the lock but remain attached to the boat by its tether of nylon webbing, but it would take time to retrieve it from the river and slide it into the lock. I needed both oars now. Worse, during the mere act of retrieving it I could lose the whole show.

As the sky appeared again, I let the oar go. Time dragged as I crushed my eyes shut to squeeze the water out of them. My gut informed me that we were moving downhill and downstream, the right combination. But as I looked to the left, the Slate Creek Diagonal seemed to ski across the river toward us. Actually we were surfing toward it. Normally at this point I would pivot the boat to aim the bow directly at that monster boat-flipper and then pull on both oars to ferry us away toward the right. But while I could still pivot with one oar, rowing with only one is impossible, and if I pivoted now I would be setting myself up to flip somewhere in the wave train leading into Crystal Hole. Suddenly, carrying all our passengers with the river at flood stage now seemed a lot less intelligent than when I had okayed it moments before.

Why was I even still doing this? I could not stop this nagging question from intruding. It was a stupid time to wonder about it, but I could not stop myself any more than I could stop this boat. I no longer had to row a boat to earn a living. I had other skills, more important skills. Why risk losing my relationship with my family, the most important thing in my life? Finally I shook this growing demon and jerked my mind back to the immediate problem.

I knew that fishing for the lost oar at this point would be the wrong move. I saw that I had rowed us so far to the right and had gained such great momentum by the time we slammed into the diagonal leading into the New Wave that now we would not be carried all the way left into the Slate Creek Diagonal. So now I struggled with the right oar to straighten the half-swamped boat to meet the waves head on.

Smash! We exploded through another wave, then through a third. The boat rose up and then dived down huge roller-coaster waves as if auditioning for a commercial for Colorado white water. I was the only one aboard uncertain about the outcome of all this.

I tried to guess whether the next wave was big enough to suck me out of the boat if I made a stab at retrieving my left oar. Crystal Hole was only a couple of waves away. I wanted both oars in the worst way. What the hell. I shipped my right oar and dove for the left.

I grabbed it on the first try and slammed myself back down on the rowing seat as the fourth wave exploded over the bow. I fitted the blade into the oarlock and shoved. It bounced right back out. I glanced ahead to see Crystal Hole galloping up toward us from the left. This was ridiculous. I fitted the oar blade again. It finally slid home. I pulled the handle of my right oar out from under my knee. Now I could row.

So what? It was too late—for Crystal Hole, anyway. We were already there but, fortunately, immediately right—on its shoulder. I glanced into its thundering collapse as it recycled down from an explosion. The past three seconds had been like being forced to stand still and watch a rhino charge you and miss by two feet. The only reason we were not in the maw of Crystal Hole now was because I had rowed so hard, eons ago, during my entry off the tongue. Entry is everything.

But we were not out of the woods yet. I pivoted the bow left and hauled on the oars to ferry us clear of the submerged Rock Garden. I pulled stroke after stroke, hating the sensation that the river had turned to molasses. Meanwhile I started wondering about the four boats entering behind us.

I scanned upstream for the first time and yelled, "John, will you grab that bucket and bail? This thing weighs an extra ton."

"Got it," he assured me as he reached to open the gate of the carabiner securing the five-gallon bucket behind the poop deck.

Then I added to those in front, "We're almost out of the woods now. Check upstream and yell if you see swimmers."

All three quickly turned to look behind us. They looked like they had just been dragged from a well.

One after another the boats behind us appeared on the tongue and then vanished into the white water. The yellow synthetic fabric of the Domars peek-a-booed through the explosions, then surfed to the left and vanished into the New Wave. Each reappeared in the wave train leading directly into Crystal Hole. I winced. Then I rubbernecked all the way down lower Crystal past the Rock Garden and hoped that I would not need to rescue a boat—or two.

One after another, each boat surfaced like a submarine from the giant standing wave below Crystal Hole. I could not believe my eyes when I saw that each one was right side up. I later received a blow-by-blow account from an observer on another trip that had arrived just behind us to scout Crystal. He said that our first boat (mine) was the only boat not to end up in Crystal Hole. It had looked so bad out there, in fact, that his entire crew started talking about portaging.

One boat, Bruce Helin's, crashed into the standing wave of Crystal Hole, jerked to a halt, shuddered, and almost buckled. The wave buried Bruce's passengers in the bow and also washed him off his seat and tossed him to the rear and into the river. Just as he was about to lose all contact with the boat, Ellie, his sole passenger in the stern, grabbed him. She weighed only about half what he did, but she delayed his exit long enough for him to grab a rigging strap and haul himself back to the rowing seat. But before he could tackle rowing again he had to locate and retrieve his oars. Meanwhile, the immense hydraulic jump of Crystal shook the boat like a lottery ball before the drawing. Luckily the wave collapsed slightly for an instant, and the boat suddenly erupted from the hole and shot toward the left cliff bordering the rapid.

"That was wild! Better than insider trading!" Wilson exclaimed as our trainee boat crashed into the cliff immediately left of Crystal Hole, where equipment was stripped from it and scattered into the river. I studied the mayhem as the frenzied trainees were carried toward the Rock Garden.

"A piece of cake!" John added from the back of the boat as he stood up to stretch his back. The trainees now fought their way to the channel left of the Rock Garden. All five boats and all the passengers had made it. It seemed like a miracle. Only a few oars and ammo boxes and other minor impedimenta now bobbed loose in the tail waves of Crystal. Relief washed over me like one of those waves. My tunnel vision finally evaporated.

I don't know why we never see it while we are scouting this rapid (well, I guess I do), but the landscape surrounding Crystal is unearthly in its beauty. Massive cliffs of ancient gray schist jut jagged and hard from the river's edge and soar a thousand feet to where they are truncated and topped by orderly-looking beds of ledgy Tapeats Sandstone topped by 4,000 feet of Paleozoic cliffs. The place is immense. In the cliffs of schist surrounding us are gigantic dikes of pink Zoroaster Granite metamorphosed into frozen veins zigzagging across the canyon walls like the petrified circulatory system of an embryonic planet. This place seems like the organic roots of the world, a place humans were not meant to understand. A deep, secret cosmic workshop where gigantic forces create the very stuff that makes up our island spinning through space. Yet, somehow, despite the violence of

Crystal, it seems a place of peace. Now, below it in Thank God Eddy, I could see what had been obscured by the demands of survival: this place is beautiful beyond my wildest fantasy.

In fact, despite its having inspired fear since 1967, and terror since 1983, as of 1990 Crystal had claimed only four river runners as sacrifices. Ironically, a midair collision above it on June 17, 1986, between a tour helicopter carrying five people and a twin-engine tour plane carrying twenty others claimed all twenty-five lives a mile from this rapid. Tire skid marks on top of the chopper hint that the fixed-wing had been banking hard when it collided with it above Tuna Canyon, but the cause of the collision remains a mystery. Had they flown here to ooh and ah over Crystal? This was the fourteenth crash in ten years in the Canyon, and it alone exceeded the death toll of sixteen river runners in all of Grand Canyon during the entire century between 1889 and 1990.

"John," I said over the roar, "thanks for the bailing job. Could you stow that bucket? Then loosen one of those straps suspending the drop bag and grab some beers. It's time for an ABC party."

"You got it," he said, dropping the bucket like a hot rock and reaching for a strap.

"*What* is an ABC party?" Susan turned and demanded from the front. She looked like a shower ad for Dial soap.

"Alive Below Crystal."

Here in Thank God Eddy I was as far away from Crystal Rapid as I could possibly be—two weeks away. Our rowing trips were two weeks long; hence at this point every mile from here on would take me that much closer to running Crystal again.

I stomped on the brake pedal and down-shifted for the first slow curve in a hundred and thirty miles, the curve just south of Navajo Bridge. I was suddenly aware of passing several Navajo roadside stands selling beads and jewelry and of looking through a windshield at a sign advising "Fifteen Miles per Hour," and of having a steering wheel in my hands. Driving the empty desert for the past couple of hours had untethered my mind. I shook loose my reverie about Crystal. It was not easy. Memory lane is a seductive road.

It was no coincidence that Crystal had started haunting me before Navajo Bridge. In a few seconds we would get our first view of the river and its flow. Although we would not arrive at Crystal for a week, the flow here would give us an idea of what we would face there later. I don't know why I was concerned about it already. Well, actually, I do. We all were curious. The flow of the Colorado dictates what kind of trip we could run.

For me this river had become more than a river. It was both the ultimate freedom and a prison. Was my wondering about the flow merely one more symptom of my increasing self-questioning on whether it made sense any more for me to keep coming back here?

I steered the van off a 470-foot cliff onto the narrow bridge. The rattle of the van and the hum of its tires echoed off the steel railings. Far below us the Colorado flowed by serenely, a translucent green snake crawling blindly to Mexico. Most of the water down there would not complete that journey. Instead, it would end up saturating some salt-loaded field of alfalfa. At any rate, the flow here seemed to be in the mid-teens; Crystal would be a cinch.

From the shotgun seat, Fabry, just arrived from Sonora to work his first trip of the year, stared down at the river as if at a vision of the Virgin and said, "There she is."

What was the magnetism? We returned every spring like migratory animals. Neither of us could stay away. I glanced at him, still absorbed in gazing down past the steel rails into Marble Canyon at the West's most famous river. For both of us the magnetism was not just rivers. We had run dozens. I had rowed or paddled wild rivers on every major continent on this planet. The magnet was the Colorado, and not just the Colorado—the Colorado here, in Grand Canyon. This place had become holy. It had nothing to do with the Grand Canyon being one of the Seven Natural Wonders of the World and a World Heritage Site. This was sacred ground for far more personal reasons.

A bulky Winnebago motor home lumbered onto the narrow bridge—it was only eighteen feet wide—from the north and crept toward us at about ten miles per hour. Both riders in front craned their necks to scan the depths below us. I imagined their conversation: "What do you suppose it's like down there, Martha?"

I knew.

2
An Agreeably Confused Appearance

Afew hundred yards north of Navajo Bridge, silhouetted against the red rampart of the Vermilion Cliffs, sits Marble Canyon Lodge: a motel, restaurant, gas station, post office, and gift shop loaded with Indian jewelry, regional books, and cheap camping and fishing equipment for river runners and trout fishermen. Even though the air-conditioned complex is built of native sandstone, it appears more like a mirage in the shimmering heat than a structure that belongs here. Detroit's (and Japan's) best steel horses are nosed up to its walls as if seeking relief from the desert heat.

Beyond the edge of this commercial oasis the desert stretches south, invisibly leap-frogging the hidden crack containing the Colorado River, then spreading beyond for vast, empty miles of rolling plateau. From the Vermilion Cliffs northwest of here to the Echo Cliffs distant in the southwest, there is little sign that people have visited this part of the planet. Scant desert shrubs—stunted pygmies due to a lack of water—dot the stony soil at intervals. Hot winds gust in from the distant southern horizon as if someone had unplugged the atmosphere behind us to let it all escape through some rent in the fabric of hyperspace. This tidal wave of hot air seems a warning sent via primeval telegraph that something is about to happen. But it is a false message. Here winds like this sometimes blow for a week.

Without stopping at Marble Canyon, I steered the van north and right off Highway 89A, upriver, to drive down the five-mile road to Lee's Ferry. We had two weeks of food to unload and five boats to rig.

This road along the foot of the towering wall of Vermilion Cliffs is an excursion into a land where the law of gravity seems to have been suspended. This ancient route is guarded by giant stone toadstools and hoodoos—huge rusty sandstone boulders perched precariously on unbelievably slender pedestals of fragile purple shale growing from barren, packed gravel. Jostle one and be crushed like a gnat. Steep red buttes jut into blue sky like crumbling fortresses of some megalithic elder world. They beg to be climbed. Deep canyons gnaw into the cliffs on the left, then dive into Marble Canyon on the right. The whole place seems too big, too abrupt, and too strange to be real, like a gigantic movie set built on an unlimited budget or scenery stolen from a Roadrunner cartoon. It requires only an empty crate reading "Acme Rocket Shoes" and a faint "beep, beep" in the distance to make one nervous and watchful for a cliff precipice and Wile E. Coyote plummeting from the sky to puncture the pavement in front of the windshield.

Although today the traffic here is limited to river runners heading downstream and fishermen heading upstream in quest of record trout from the chilly Colorado's new planted-trout fishery, we were speeding down one of the most ancient routes on the North American continent. This route owes its existence to the ancient Echo Cliffs Monocline crossing the Colorado upstream and slicing into bedrock to create a weak zone. Countless grinding flash floods during thousands of millennia have funneled along this weak zone to carve a rare access route across North America's longest canyon system. Roughly 11,000 years ago, when the last continental ice sheets were melting and receding into Canada, Clovis hunters pursuing the vanishing Pleistocene megafauna had trekked this way to the mouth of the Paria River to ford the Colorado in search of the shrinking herds of imperial mammoth and giant bison. At the same time, in western Grand Canyon, these hunters wiped out the last of the ponderous Shasta ground sloths with their deadly spear throwers, called atlatls. At that time, much of the Colorado Plateau was lush grassland and pine forest, and except for the ford at the now-submerged Crossing of the Fathers upstream from here in Glen Canyon, the Echo Cliffs Monocline at Lee's Ferry was the only easy place for people to reach the river and cross it at low water between Hite and Pierce Ferry, a distance of 440 miles. It is a safe bet that these vanished First Americans did not refer to the place as Lee's Ferry—but, then, perhaps neither should we.

In July 1776 two Spanish priests, Fray Francisco Atanasio Domínguez and Fray Vélez de Escalante, in company with Captain Bernardo Miera y Pacheco and ten mounted soldiers, departed Santa Fe in search of a northern route to California. Storms halted them in Utah, where they decided to seek a shortcut back to Santa Fe. By late October they had found the confluence of the Paria and the Colorado, guessing wrongly that it was the Ute Ford. Two men bundled their clothes on their heads and tried swimming the river, but they lost their clothes and barely won against the strong current to regain the north bank. Then they tried crude rafts, but these also failed. Despite their predicament, Escalante wrote of the Echo Cliffs Monocline— and future Lee's Ferry—in his journal, "It has an agreeably confused appearance."

It was now November, and the weather was anything but agreeable. It was foul, and food was low. The Spaniards knew a route must exist upstream. After a hellish struggle to lead their horses through wind-blown dunes while climbing up 2,000 feet of precipitous and crumbling slopes in the Echo Cliffs, they trekked for five days through deep red sand, weaving in and out of sinuous arroyos carved in the sandstone and peering over the rim of Glen Canyon in search of a route to cross the Colorado. Finally, in the shadow of Gunsite Butte, they found a canyon that dropped 500 feet to a promising stretch of river. The men chipped notches in the steep sandstone of what they called Padre Canyon to give the horses footing. They rode across the wide Colorado, then climbed out Navajo Canyon, following foot holes pecked into the steep stone by the Anasazi seven centuries earlier. This was the Ute Ford, renamed the Crossing of the Fathers because of them.

So why is it Lee's Ferry? In 1858, during the rise of the Mormon empire in Utah, the State of Deseret, Brigham Young sent his premier scout, Jacob Hamblin, on a mission south to proselytize the Hopi, whom he had heard were peaceful farmers living in towns. Hamblin hired a Paiute named Naraguts to guide him to the Ute Ford. But Naraguts inadvertently led them to the mouth of the Paria, and like Domínguez and Escalante, Hamblin's party was forced to ascend northward from the Paria and to rediscover their crossing.

Despite having located the only two good routes across the Colorado, Hamblin failed with the Hopi. This oldest civilization in North America (they were descendents of the Basketmaker II people, who dated to at least 1000 B.C.) wanted nothing to do with the Latter-day Saints' religion or any Christian religion. Hamblin trekked several more times into Arizona over the next decade, but not until 1864 did he cross the Colorado near the mouth of the Paria. Hamblin then recommended it to Brigham Young as the best site for a ferry to open northern Arizona to Mormon colonization. Seven years later Young acted on Hamblin's suggestion, and the man Young sent to build the ferry was John D. Lee.

Lee's mission had been no reward—except as a reprieve from execution. Colonel Albert Sidney Johnston's 1857 invasion of 2,500 soldiers into Utah to depose Brigham Young had raised the Mormon sense of persecution by Americans outside their religion to a fever, and this fever prompted atrocities. John D. Lee had been a kingpin in the most notorious incident. Lee was a high-ranking member of the Mormon Nauvoo Legion and the secret militia called the Sons of Dan, whose military modus operandi included dressing, acting, and killing as Indians.

In that same year, 1857, only a decade after the Mormon hegira from Illinois to Utah, a wagon train led by a man named Fancher entered southern Utah en route to California. The party insulted the Mormons: Fancher himself drove an ox he had named Brigham Young, whom he boasted he loved to whip. The Fancher train was also stupid enough to camp in Mountain Meadows west of Cedar City to fatten their oxen and other livestock preparatory to the lean, dry trek across Nevada. In September the tension exploded as the Mormons' Paiute allies attacked the Fancher train in retaliation for their having shot some Paiutes. They fought to a stalemate. The Indians' failure to wipe out the 140 Gentiles prompted John D. Lee, the local Indian agent, to be ordered to "manage" the Indians and complete the job they had botched—and to cover it up.

The cover-up team consisted of the Paiutes and about fifty Mormon men, including Lee. The Mormons persuaded the men of the Fancher train to surrender their arms in exchange for halting all hostilities. (Bear in mind that they knew the time they were losing under siege would soon strand them foodless east of the Sierra Nevada; their losses in livestock made this even worse.) Then the Mormons separated the disarmed men from their women and children and shot the men to death where they stood. The shots signaled waiting Paiutes to slaughter the unsuspecting women and most of the children. Only the eighteen smallest children survived.

Wiping out every member of the Fancher train old enough to bear witness did not preserve the secret, however, despite a blood oath sworn by all—too many men had participated. Word leaked out in 1858. In 1859 the U.S. Army arrived to investigate the massacre. A warrant for Lee's arrest failed because fellow Mormons hid him, despite his formal excommunication from the church. When in 1871 Brigham Young finally ordered Lee on a mission to build a ferry across the Colorado at the head of Grand Canyon, he was sending Lee to Deseret's most distant and isolated frontier for his own protection.

Accompanied by two of his nineteen wives and several children, Lee homesteaded the mouth of the Paria. A year later he built a proper ferry, but the Navajo soon drove Lee back into Utah, where he was finally arrested in 1874. After two long trials, a Mormon jury found Lee guilty and sentenced him to death by firing squad. He was shot in 1877—the only participant of the Mountain Meadows Massacre to be tried or punished. Lee left behind fifty-seven children. In 1950, Juanita Brooks, a Mormon, wrote *The Mountain Meadows Massacre*, revealing the incident in detail and disclosing the very minor part Lee had played in the killing. Eleven years later, Mormon officials reinstated him posthumously in the church.

I drove across the bridge spanning the trickle flowing down the narrow sandstone canyon of the Paria. This is a great three-day hike—as long as you avoid the summer monsoons. Flash floods in the narrows scour it wall to wall.

A half mile beyond the Paria I steered the van onto the launch ramp. Our white GMC flatbed truck had left Flagstaff ahead of us and was now backed down to the river's edge. We weren't the only professionals on the loading ramp. Tanned boatmen wearing baseball caps and shorts with pliers and knives dangling from their belts were loading gear and supplies on five huge, motorized pontoon boats. Two belonged to Ted Hatch, and three to Georgie White, the legendary 77-year-old woman outfitter. Our crew of six started hauling our equipment off the flatbed. In two hours we'd be rigged and ready to row for two weeks.

Things have changed at Lee's Ferry. For half a century after John D. Lee was executed, Lee's Ferry remained the most critical part of the main artery south into Arizona. The ferry operated until 1928, closing only after a blunder led to the boat's nos-ing under in strong current, flipping, and then snapping loose and vanishing down Marble Canyon. Both the passengers and the operator were swept downstream and drowned in the silty flow while their Model T sank like a stone and the operator's wife watched in horror from the far shore. Lee's Ferry died with them.

The opening of Navajo Bridge at Marble Canyon in 1929 made Lee's Ferry obsolete, a sand-blown delta punctuated by a few crumbling stone buildings. But the rise of river running in the 1960s—and simultaneously Glen Canyon Dam's conversion of the Colorado from a warm, silt-laden river to a clear, cold trout stream—put Lee's Ferry back on the map.

Although the Colorado in Grand Canyon has not been tamed by Glen Canyon Dam fifteen miles upstream from here, it is controlled—at least most of the time. Electricity varies in price depending upon when it is available, being twice as costly during working hours as during the night. So the brokers of the Western Area Power Administration (WAPA) instruct the engineers of the Bureau of Reclamation to spill the greatest flows through the turbines between 7:00 A.M. and 3:00 P.M. and again in the evening. At night they drop it drastically to conserve water for the next peak demand—and peak sales price. This "peaking power" strategy may vary hourly and normally causes a double "tide" daily, often shifting the flow abruptly from 3,000 to 30,000 cfs, or vice versa. It also gives a new twist to the truest maxim of the Southwest: "Water is power, and power is money."

Just one side effect of all this is the settling of roughly half a million tons of silt every day in Lake Powell behind Glen Canyon Dam and the transformation of the river below it into the Southwest's blue ribbon trout fishery. Lee's Ferry is the only launch point for motorized fishing boats to ply the calm waters of the Glen Canyon's remaining fifteen miles in the quest for monstrous rainbow trout. Here chemical nutrients in the river are assimilated by an jungle of cladophoran algae, which is grazed upon by the larvae of buffalo gnats and midges, upon which the trout depend. The best fishing seems to be near the dam. Below the Little Colorado (at Mile 61.5 from Lee's Ferry) the fishing fades.

"I hope this trip's got some women on it," Fabry said tiredly. "I need to get away from my situation in Sonora."

Once we had rigged our eighteen-foot boats, we rowed them 200 yards upstream into the mouth of Glen Canyon and tied them off the sterns of Ted Hatch's thirty-three-foot motor rigs moored in deep water there to prevent them from being stranded in the morning by the bureau's parsimonious flow. The Hatch guides were friends. Then we drove to the Vermilion Cliffs Restaurant.

The crack of Jimbo's cue on the one pool table in Vermilion Cliffs reverberated like a pistol shot across the small log dining room and bar. Jimbo and our second trainee, Jim Sutton, held the table. The room was jammed. All six dining tables were pinned to the floor by the elbows of boatmen from five companies who in the morning would be rigging or launching boats at Lee's Ferry. Half of these guides were probably wishing the same thing as Fabry.

"I thought that situation was over," I said. This was in May—his first trip of the year—and we had some catching up to do. During our years of running rivers in America and Africa, we had developed a rapport possible only in the cauldron of shared adversity and triumph: attacks by irascible hippo on the Omo in Ethiopia, tough rescues in icy white water tumbling from the Sierra Nevada, scorching hikes across the Arizona desert in heat that could suck half a gallon of water out of a man in an hour . . . and voracious women on soft beaches far from the society they had temporarily abandoned. I had missed him over the winter.

"It is over with Cheri," he admitted quietly. "But in the past few months I've been seeing someone else." Fabry was a handsome specimen, a candidate for the next Marlboro Man and he knew it. His modeling portfolio had already landed him fat residual checks from a television commercial for chewing gum. You could not tell he was really chomping on a huge chaw of weed.

But we'd had this conversation many times over the years. Only the names had changed. Fabry was searching for the perfect woman—one who was beautiful, great in the sack, independent, intelligent, with a good sense of humor, a self-starter, ambitious, able to provide a healthy second income, fertile, and hopefully well connected. And he had other qualifications. But as everyone knows, the perfect woman is hard to find, and as most men forget, the perfect woman is normally going to become serious only about the perfect man.

Like many people, Fabry was more than he appeared. His master's degree in cardiac medicine had been inspired by his failed attempt at CPR on his father during a heart attack. Fabry now battled his father's killer nine months a year by teaching cardiac rehabilitation in a California community college. In fact, he had flown from school last night after posting his grades, and I had picked him up at the Flagstaff airport.

"So," I asked to change the subject, "four trips back to back?"

"Yeah," he said, rubbing his neck as if to erase a judo chop. He seemed half proud at his backbreaking schedule and half dismayed.

I shook my head. "Fifty-six more days without a break. You should get some sleep tonight." During the middle of each of the past three years I had worked sixty river days without a day off, but I usually had taken a couple of days off before starting them. To survive this you had to pace yourself.

"It's catching up with me," he said. I could see that it was.

(Neither of us knew it now, but he would not complete these four trips. On his third, at the confluence of the Colorado and the Little Colorado, his sudden collapse into an acute fever would force me to radio for a helicopter to evacuate him. The doctors at the Flagstaff Medical Center would diagnose him with a systemic infection from cellulitis and with a punctured and collapsed right lung.)

Under an ink-black sky twinkling with an astronomer's obsession, I jumped from the half-inflated military bridge pontoons of Hatch's rig onto my boat. I lifted the front hatch of diamond-plate aluminum and grabbed my Vietnam surplus waterproof rubber bag (called a black bag) from the drop bag beneath. Then, sitting on the 172-quart dairy cooler that doubled as my rowing seat, I unbuckled and unrolled my self-inflating Jack's Plastic Paco Pad—the best friend of my lower back—and whipped it across the rear deck. Although we don't talk about it much, most boatmen consider their boats sacrosanct refuges from the statusless existence on shore. A boat is a good place to keep your gear organized and is the best place to sleep, buffered from the desert heat by the water and distant from the scorpions that stalk the beaches at night.

These Domar riverboats were excellent. In the 1950s and 1960s, cheap military surplus had boosted whitewater rafting from an eccentricity to a lucrative commercial enterprise. At first, U.S. Army ten-man assault rafts from World War II, available for a couple hundred dollars each, were state-of-the-art technology. But demand outgrew supply, and as boats wore out or were destroyed by less-skilled boatmen, no new ten-mans were to be had. The price of new assault rafts sky-rocketed so high that professional outfitters (a thrifty breed at best) choked at the mere mention of buying them to replace the dead ones. But as the American whitewater touring industry grew to more than 700 outfitters, some of whom owned a hundred disintegrating boats and needed replacements plus additional boats for expansion, several manufacturers of inflatable sport boats jumped into the whitewater arena. As guides, we had rowed boats built by every major manufacturer, many of them lemons. Now, finally, we had Italian-made, inflatable Domars, floating Maseratis, the best boats available: at eighteen feet long and eight wide, and with identical bow and stern, they were versatile, fast, responsive, and photogenic. And good bedrooms.

Our bedrooms were now floating almost directly over the Echo Cliffs Monocline. Lee's Ferry is bracketed upstream by the final sculpted sandstone cliffs of Glen Canyon, immense and sinuous and, until twenty-five years ago, the protective vault for roughly 170 miles of river canyon generally agreed to be the most serene and beautiful on the Colorado.

Glen Canyon Dam converted Glen Canyon into a submarine canyon, a vast inland lake—the world's second longest reservoir—stretching 186 miles up the Colorado past the head of Glen Canyon to also submerge Narrow Canyon and the foot of Cataract Canyon. Scores of tributary canyons, once miniature Shangri-las of the vanished Anasazi Indians, were also flooded by the dam and now provide Lake Powell with 1,960 miles of shoreline—more than the entire West Coast of the United States. But now, in the clean starlight, Glen Canyon regained its former grandeur before the coming of the Bureau of Reclamation. The view was stunning, monumental against the black of outer space, like the fantastic ruins of a long-dead civilization slowly eroding to dust on a lifeless planet millions of miles from Earth.

Here the flow was so serene that only the faint ripples of reflected starlight revealed that it moved at all. My boat was calm. A waterbed is a rougher ride. But the night was alive. In the lush vegetation fifty feet away, a legion of crickets chirped as if this were their last night to find a mate. Guided by sonar, small brown bats swooped, climbed, dived, then turned on a dime in impossible gyrations to scour invisible gnats and midges from the air. One little flying mammal winged so close that I felt the wind of its passing on my cheek as it blotted out the stars. No danger of a collision though—sonar.

I studied Cygnus the Swan, the Northern Cross, then one by one I ticked off the constellations surrounding it: Draco (the

Dragon), Lyra (the Lyre), Delphinus (the Dolphin), Aquila (the Eagle), Cepheus (the King), and even tiny Sagitta (the Arrow). For some reason these constellations twinkling with such diamond purity seemed like friends.

A warm breeze stole heat from a stone landscape that had absorbed infrared rays all day and flowed across the boat to displace the cool air rising from the river. On a night like this you could do things.

I glanced downstream toward the Vermilion Cliffs to the right and the Echo Cliffs to the left, and toward the reincarnation of the tireless Colorado picking up speed between them. My heart beat faster. Instead of probing a vanished past, I now itched with curiosity about what this trip might bring. The Canyon itself offered many challenges but only limited surprises. People create surprises. Without doubt this trip would have its moments: high, low, memorable, unpredictable. I wondered, would the perfect woman show up for Fabry? In my wildest dreams I would not have imagined the woman who did show up.

Over time I have come to think of each of these trips as a benign plane crash. All that the twenty-two of us who would be depending on one another really had in common was a desire to go to the same place and have fun doing it. Here we were, cast into the desert wilderness, but luckily with no injuries and with enough food, equipment, and inflatable rafts to survive. The captain of the plane (me) persuades everyone not only that everything will work out all right but that if we approach this disaster with a positive attitude, it could be the experience of a lifetime. After all, we are healthy and well equipped. What else is important? It will take us two weeks to make our way to the Hualapai Reservation, where one of the Indians can rescue us in the tribal school bus. By then the Canyon will have welded us together.

Tomorrow was predictable. At midmorning sixteen strangers would spill from our bus, slick with lotions and spotless in white clothes over stylish new swimsuits. They would smile in anticipation behind sunglasses and under funny hats—or worse, no hats at all. They would drag out possessions selected in agony, half of which would lie unused in the bottoms of their black bags. We would greet them and fit them into life jackets. They would take one step into forty-nine-degree water to climb into our boats—bringing with them God-only-knows what expectations. I would orient them with a safety talk on how to avoid the most common dangers: dehydration, sunburn, loose boulders on the hikes, going bare-footed, and hypothermia while running rapids in early morning. I would explain protocols for whitewater emergencies, plus camp tips. And worst of all for some people, I would warn them about the creepy-crawly menaces in camp at night: red ants, scorpions, rattlesnakes, and black widow spiders. By the time I finished, some people would be wondering how they could have made the idiotic decision to come on a trip like this. Then, as a fledgling but uncertain family, we would float into the most impressive canyon in the world.

All that was predictable. What I could not help wondering was what would happen after we launched. The Canyon strips people of their studied personas to reveal the true individuals beneath. Psychotherapists, eat your hearts out.

This canyon shoves human concerns into perspective, whittling personal problems down to the toothpicks they really are. Maybe that's why I can't stay away. I'm sure that's why. This place does something to people, something good, even the challenges. In fact, from a human perspective, it is not the challenges themselves that fascinate us so much but how people riding the river into this canyon—the few in the past and the more than 21,000 a year today—*react* to them.

My eye caught the blazing trail of a meteorite flashing across the crowded stars of the Milky Way in an arc of fire toward its death in the Painted Desert. Things seem simpler here, more elemental.

They are.

3

Into the Canyon

A strong gust of hot wind blasted against my back as if some demon had unlocked the exit from hell, and instantly my oars froze in the river. My boat refused to budge downstream. The wind robbed me of my rhythm. My strokes ground to a halt. The ton of boat beneath me, shoved by the gale, started drifting against the current and back upstream toward Lee's Ferry. I heard myself grunt as I pulled even harder, refusing to allow the wind to stop us.

As my boat regained its snaillike momentum and slapped against the wind waves, I glanced upstream at the four boats struggling to follow me against the river of hot air funneling up Marble Canyon. No one could confuse this grinding, machine-like rowing by each boatman with pleasant exercise, but we had no choice. We had to row these miles or end up camping on a pile of boulders blasted by sand.

My lower back felt the strain. It was at times like this, when the rowing was brutal, that I most wondered whether it still made sense for me to remain a fugitive from the twentieth century. Sometimes I felt as if I were hiding in the Painted Desert, dragging my feet—or oar blades—while the rest of the world stumbled headlong into the next century.

I pulled another hard stroke to correct my angle as a new gust of wind torqued the boat. The river seemed like molasses.

Here at Mile 7 the Colorado was nearly a hundred yards wide between cliffs of pink-stained limestone, buff sandstone, and deep red shale already 600 feet high. Upstream and downstream the river seemed an elongated lake of flat water rippled by wind, and even though we were only two hours downstream from Lee's Ferry, the Canyon already seemed impossible to escape except by helicopter. These cliffs seem inescapable to every new visitor. Francisco Tomás Garcés, the Franciscan friar who, during his solitary search for pagan souls in 1776, became the first European explorer of the southern Grand Canyon region and the first consistently to refer to its river as the Rio Colorado, called this canyon a "*calabozo* [prison] of cliffs and canyons." It's easy to see why.

I pulled again. The wind could not last forever, but this knowledge did not persuade my heavy boat to respond any less sluggishly to my strokes for correction. It felt as if I were dragging a sea anchor.

From habit, my eyes scanned the cliffs of Kaibab and Toroweap formations atop the Coconino Sandstone. This was classic early Marble Canyon: green river, red walls, blue sky. Clean and simple—but enormous.

Across this windy blue sky flitted an iridescent cloud of violet-green swallows, changing course so abruptly to gobble buffalo gnats that they seemed to be giant insects instead of birds. This species is ubiquitous along western rivers. Here in the Canyon the swallows were in turn preyed upon by peregrine falcons diving from the heights at a hundred miles an hour.

I have witnessed several blitz attacks by this endangered species (which, after serious declines due to pesticides and captures by unscrupulous falconers worldwide, has made a local comeback to build North America's largest breeding population here in the inaccessible fastness of the Canyon). The peregrine banks close along the contours of a cliff face or spots its prey from a perch, "waiting on" nearly invisible against a cliff, then plunges in a stoop at blurring speed, either striking the swallow from above with enough momentum to literally knock it from the sky, or, after a chase, spinning beneath it to strike from below. The falcon may bind to its prey for the blink of an eye, then follow the stricken bird to alight with it on terra firma or swoop down to pluck it from the river. Occasionally the falcon flies close to shore, often only inches above the rock, using the same strategy to avoid detection as a Stealth bomber does to surprise its prey.

Peregrines are so impressive that the swallows seem outclassed, especially when the falcons pursue in pairs. But natural selection has equipped swallows with maneuverability, allowing them to prey on insects and to escape falcons (most of the time). Even ducks and pigeons escape them.

The distant roar of Badger Rapid (Mile 8) drifted to the boat. The accelerating current would soon allow me to take a break without being blown back upstream. I congratulated myself now for not stopping at 6-Mile Wash. The wind was so brutal that I had been tempted to call it quits and camp there. I have never camped there before, nor heard of a trip that has. It is not a good camp at all, and it would have put us behind schedule. Despite the ignominy of surrendering to the wind, though, the crew would have been happy to stop.

I glanced toward shore. Hot wind blasted the scant tamarisk into convulsions. The only refuges from these sandstorms were out here in midriver or well up a side canyon. But staying here required constant rowing, and I was already regretting the fifteen cases of beer in the drop bag under my rear deck; their weight bulged the floor downward and submerged the stern so much that I felt as though I were rowing a D-9 Caterpillar through sand.

Sometime before midnight the wind would die down. There was no sense now in trying to row beyond the excellent camp below Badger Rapid. Probably the wind would not pick up again until late tomorrow morning. During the three or four hours before it did, we could push downstream another twenty miles. Although Badger Rapid was still a few hundred yards distant, a strong blast from Satan's furnace abruptly escalated its roar. A reassuring sound. The sound of camp. The sound of emancipated back muscles.

Badger is the first rapid in the Canyon. On the so-called western scale of one to ten, it rates a five or six. If you were asleep at the wheel, you could flip a boat in it. It still happens, although it shouldn't.

Most rapids here are named after the tributary canyon that created them by funneling a gigantic flash flood of rock slurry into the Colorado and dumping an alluvial fan of boulders across it. These natural "dams" cause a sudden descent. They also constrict the channel and complicate the topography of the riverbed. These factors—gradient, constriction, and topography—combined with the volume of river flowing over them, make rapids what they are.

But as soon as a rapid was born here, the river started tearing it apart. Because the river's capacity to carry rocks increases dramatically as the current doubles, the peak flows in June used to dismantle and rearrange new rapids, generally making them easier. But the Colorado's ability to do this eroded in 1963, when Glen Canyon Dam limited annual peaks to around 30,000 cfs. Except for the freak flood of 97,200 cfs in 1983, and those of half that volume in 1984 and 1985, the bureau has choked the old Colorado so thin that it can no longer change big, new rapids much.

Anyway, I was in no hurry to run Badger. The wind allowed me to stall our entry, and the current matched it to prevent us from being blown back upstream. I tucked my oar handles under my knees to hold the blades out of the current, then I relaxed against the black bags strapped to my rear deck. Soon the other boats would reach this zone of equilibrium. Being out here suddenly felt good again. The muscles in my lower back sent thank-you cards, cards that promised thousands more miles and hundreds of thousands more strokes of the oars.

In the next twelve days we would add to today's 8 windy miles another 218. Our ultimate destination was Diamond Creek, a dozen miles upstream from Lake Mead. Except for those last dozen miles, we would row the entire river while remaining in Grand Canyon (which, measured between Lee's Ferry and the Grand Wash Cliffs before Boulder Dam was built in the 1930s, used to be 277 miles of flowing water, but today it has lost 40 miles to Lake Mead).

Two hundred and twenty-six miles of river. But not just a river. The white water here routinely disintegrates the composure of professional boatmen. The Colorado descends an average of eight feet per mile in Grand Canyon. This does not sound like much, and in fact it isn't. I have run rivers with much steeper gradients, up to twenty times steeper. The Alas River in Sumatra, for example, drops 165 feet per mile for three miles in its most difficult section. Even in the United States, river gradients exceeding 50 feet per mile are considered commercially manageable—the Tuolumne and the Forks of the Kern in California, for instance. So at first glance it may be hard to imagine how a mere 8 feet of descent per mile could create the Colorado's notorious rapids.

Inescapably, much of the river is flat water, as flat as Kansas, repeatedly and for miles at a stretch. This is one reason why some rapids are so violent. The natural dams that create these rapids pond the Colorado for miles upstream to create the flat stretches. Then abruptly, where the river pours over one of these boulder dams, it erupts into violence nearly unsurpassed on any other navigable river in North America. The overall result is dozens of sizable rapids, many of which drop fifteen feet or more, punctuating the 226 miles of mostly flat river between Lee's Ferry and Diamond Creek. The biggest, Hance, drops thirty feet. Although many rapids are easy for a veteran, some are difficult, and a few are horrendous.

The Colorado has carved such a gigantic labyrinth of badlands that it was the last major chunk of terra incognita in the continental United States to be explored. Before Glen Canyon Dam went on-line in 1963, the Colorado here used to hit an average peak at about 85,000 cfs in June. In 1921 it peaked at 200,000 cfs. River historian Dock Marston reports that the maximum flow ever recorded occurred on July 8, 1884, at 300,000 cfs. Sediments reveal that during the past 3,500 years, the river has exceeded 250,000 cfs fifteen times. At least once it topped 500,000 cfs. The Colorado still floods, but Glen Canyon Dam squelches high flows above Grand Canyon. In 1984, for instance, the inflow from Cataract Canyon into Lake Powell peaked at 148,000 cfs, but the flow here never exceeded 50,000 cfs.

In spate, however, much of the Colorado is not water. Because of the tremendous sediment load it carries from its 242,000 square miles of watershed—mostly desert sprawling across parched and friable sedimentary bedrock packed with petrified carcasses of dinosaurs scattered like peanuts in chocolate—this river long ago earned the wry endearment, "too thick to drink and too thin to plow." This was no exaggeration. In the late 1940s, before the dams were built upstream and when the Colorado exceeded 100,000 cfs, only 48 percent of what flowed past was water. The rest was sediment. The river normally transported a million tons of sediment past Lee's Ferry every couple of days. Today that sediment is piling up in Glen Canyon. The Colorado River itself is impressive, but compared to its course a mile deep in Grand Canyon, it is a mere artery of snowmelt and sediment slurry. It is the canyon that leaves us gaping in awe.

When Woodrow Wilson signed the bill establishing Grand Canyon as a national park on February 26, 1919, it was neither the first nor the biggest national park in the United States. But of all the national parks and World Heritage Sites—of all the real estate on this planet that I have seen—the Canyon wins the prize

42

for being both the most unusual and the most impressive. It spurs instant disbelief.

Slightly deeper or longer canyons do exist—Hell's Canyon in Idaho and Barranca de Cobre in northwest Mexico, for example—but none resembles Grand Canyon. Only one other planet in our solar system is known to hold a bigger canyon. Vikings I and II sent us photos of the Grand Canyon of Mars in the chaotic Valles Marineris system. The biggest section of its 3,000 miles of canyons is 360 miles across and over 4 miles deep—formed apparently by faulting and scarp recession rather than by river erosion. It dwarfs Grand Canyon but appears overall to be the result of entirely different geodynamic processes—and, of course, it lacks a river. But it did not always; one small leg of this system is nearly identical in all dimensions to Grand Canyon and also appears to have been carved by a river. On Earth, though, Grand Canyon is unique.

It still amazes me how this gigantic place is hidden. Almost anywhere atop the resistant Kaibab Limestone (the capstone of the Coconino and Colorado plateaus) and more than a hundred yards from its edge, the flat perspective and the piñons and junipers conceal the abyss. Few guess that it lies only a stone's throw away. The surprise of abruptly looking over the rim is shocking.

My first glimpse from the brink stopped me cold. I thought I would look over the edge and say, "Yep, just as I thought." Instead, air hissed into my lungs as if I were about to make a high dive; it disoriented me; it did not compute. The sudden depth, size, and complexity seemed an illusion.

The canyon stretches at least a hundred miles to the left and another hundred to the right. Cliffs, terraces, more cliffs, more terraces, and yet more cliffs and terraces repeat again and again a mile deep. It is so deep that, were you to jump into it from some spots, you could almost recite the Preamble to the Constitution before hitting bottom. It is so deep that the huge river within it is hidden from view from the rim almost everywhere. Picture a space ten miles across containing gigantic ridges and flying buttresses of red rock snaking sinuously from the walls into the invisible depths. Imagine also a gigantic, inaccessible landscape of a thousand cubic miles at your feet, populated with scores of buttes, spires, towers, mesas, mountains, and precipitous gorges, each thousands of feet high and any one of which would be the pride of Kansas or Australia, standing forever in a landscape where water is the creator and where water is so rare that it is the most precious substance to be found.

From the rim these abrupt cliffs, free-falling tributary canyons, and gigantic spires and buttes seem the product of intelligent design. The orderly, stratified limestones, sandstones, and shales within them appear deliberately chosen as the blocks for building the largest megalithic temples in the known universe. They please the eye and seem right for the job. But again you are jolted by the sheer size of everything. The pyramids of Egypt and Mexico, the Colossus of Rhodes, the temples of ancient Greece and Rome, the megalithic fortresses of the Inca—all of this world's most famous cities and temples of stone—could be placed in one small tributary canyon below you and go unnoticed for a long time. They would be pipsqueaks in a convoluted land of stone giants. Their thunder would be a whisper.

The depth of Grand Canyon is no illusion, no trick of light. The vertical distance from the North Rim at Point Imperial to the river, for instance, is 6,600 feet. Due to the southward tilt of the Colorado Plateau, the vertical distance from the river to the South Rim averages a thousand feet less than to the North Rim. The Canyon averages about ten miles across, with a maximum of over eighteen miles.

But even beyond what one expects based on its dimensions, the Canyon is intriguing. The first thing I wondered when I peeked over the edge and finally caught my wits was, What's down there? Today this gigantic hidden labyrinth that delights astronauts is hostile for humans unless they bring supplies from more generous regions of the planet. This was not always so.

Over the millennia, Indians have lived in it or made spiritual pilgrimages to it.

More compelling to me than those lost civilizations, however, is how the Canyon exudes an intangible dimension. The heat waves shimmering off the sun-baked rock seemed to sizzle and whisper secrets about time. Indeed, Grand Canyon offers the world's clearest display of planetary history. A glimpse from the rim telescopes hundreds of millions of years into a moment. It makes me feel like a gnat.

We floated now in the Marble Canyon preamble to Grand Canyon. The other four boats had closed up. Each boatman shipped his oars to lean back against black bags. I needed no poll to know that each wanted to call it quits. Fabry's strokes had already gotten sloppy from fatigue, but Danny was the one in last place. This wind was unbelievable.

Now it ripped the crests off the exploding hydraulics of the rapid, blew them upstream horizontally, and spattered us with droplets as it amplified Badger Rapid's roar. My boat glided onto the top of the tongue, funneling into the deepest part of the rapid. In a few seconds we would speed up and pound into the big waves, and the wind would blast every molecule of splashed water across the boat as if from a fire hose. Hot winds like this have evaporated the upper foot of warm rivers like the San Juan, a tributary of the Colorado, in a day. But now the current finally gained the upper hand against the wind and accelerated us down the tongue. It was time for action.

I pivoted the bow ninety degrees toward the right shore and pulled on the oars to escape being dragged into the nasty hole just right of center. Badger was an easy rapid at 7,000 cfs, I reminded myself, a piece of cake. But the wind now shoved against the entire length of my boat, sailing it to the right. I stroked harder on the oars and felt mild annoyance at how the river seemed like molasses. Finally my floating D-9 Caterpillar plowed toward the left side. I pulled one last stroke to the left and then, at the last instant, pivoted ninety degrees again to face directly downstream.

The first small wave spread a sheet of frigid water across the boat. The next was bigger and colder. The third blasted over us. I pivoted the boat slightly right, then left, to hit each standing wave perfectly straight—the best way to avoid stalling and also to keep swamping to a minimum. Despite my maneuvers, the wind smothered us repeatedly with airborne Colorado, *cold* Colorado, courtesy of the Bureau of Reclamation architects who positioned the feeder penstocks for the turbines in Glen Canyon Dam 250 feet below the surface of the reservoir, where the water is dense and frigid. The old Colorado used to reach eighty degrees in midsummer. Heaven.

Although I say the Badger is a fairly easy rapid, not everyone agrees, and certainly the early rivermen here would not have. Just the opposite—Badger scared the living daylights out of them. In fact, the last rapid in all of Grand Canyon to be run was the second one on the river, Soap Creek Rapid at Mile 11, and this was partly because of Badger. The first ten trips to enter Marble Canyon hated—and portaged—Soap Creek Rapid because Badger Rapid had been so scary. Those few who ran Badger were thrashed so severely that the roar of Soap Creek Rapid stopped them cold with attacks of better judgment. The first nine trips, from John Wesley Powell's first descent in 1869 to the Stone-Galloway trip forty years later, hauled their heavy loads around this rapid rather than risk running it.

The pioneers who finally "stood up" to the challenge of Soap Creek were the Kolb brothers. In November 1911, Ellsworth and Emery Kolb rowed their oak boats past Lee's Ferry on low water and ran Badger Rapid. Emery miscalculated, hit a rock, and almost flipped here. An hour later they stopped to scout Soap Creek. For Ellsworth, the fact that all previous trips had groped their way around this rapid seemed the best reason to run it. His bid for a first descent provides a moral that thousands of boatmen since have relearned the hard way.

In his excellent book *Through the Grand Canyon from Wyoming to Mexico*, Ellsworth admitted that before even seeing Soap Creek he wanted to push the envelope of whitewater boating by running it—he wanted a first, and he wanted it on film. Emery's movies of him running white water in the Canyon, in fact, would be another first. But Emery was dubious. They were on their own; a serious accident could spell disaster. But Ellsworth, older by four years and possessing thirty pounds more muscle, insisted. So Ellsworth rowed out, picked his entry, was accelerated down the rapid, then crashed to a dead stop against a rock (a boulder no longer visible today). The boat lurched up on its side and ejected Ellsworth into the river. November is a cold month here. The water was freezing. Luckily, the boat spun off the rock. Ellsworth clawed his way back into the cockpit in time to row to shore at the bottom of the rapid.

A quick assessment of the damage by an anxious Emery, by far the more conservative of the two, revealed that some of their photographic plates and cameras had gotten wet. One of the main reasons for this journey was to capture Grand Canyon—and whitewater boating—on motion picture film for the first time. Emery was irritated with Ellsworth's disregard for this primary goal. They had spent ten years at Grand Canyon's south rim dreaming, saving, planning, and preparing for this trip. Emery decided to portage his boat—as their predecessors had done.

Ellsworth admits in his book to having a gnawing feeling that he knew exactly what he had done wrong and that all he needed was a second chance—and another boat—to prove it. Although he did not describe it, the two brothers probably stared at the rapid and argued. In the end, Emery acquiesced again but recommended that Ellsworth wait until morning. But Ellsworth thought he had it wired and was loathe to wait all night for something he could pull off in a few minutes.

He studied the rapid one last time, identified a cheat route to avoid the worst of it (again, this "worst" no longer exists), then rowed out. But he missed his second entry too, ran big waves he was planning to avoid, and flipped. He almost drowned as he drifted down the rapid trying to cling to the bottom of his brother's boat while being pummeled and thrashed by the cold waves. Meanwhile, Emery sprinted down the shore, jumped into Ellsworth's boat, and rowed it out to rescue him and the boat, now upside down and leaking water into undesired places. I would love to have eavesdropped on their "conversation" that night at their campfire.

The Kolbs rowed to Phantom Ranch and hiked up to their home in Grand Canyon Village, where they took a month off to develop film. Emery's wife had developed appendicitis, so he took her to Los Angeles. Then, after hiking back down to Phantom through a couple of feet of snow, the Kolbs continued down Grand Canyon. Despite more flips, mayhem, and portages over ice-covered boulders, they successfully ran the rest on low water. The brothers edited a movie of their Canyon footage, which Emery showed at the South Rim from 1915 until his death in 1976. It was the longest-running commercial film in history.

But they had not made what most boatmen consider a successful run through Soap Creek. In fact, eleven years later, when Emery was head boatman for a U.S. Geological Survey trip, he was still so steamed about Ellsworth's recklessness that he vetoed anyone running Soap Creek Rapid, even though it was flowing at such an easy level that it had almost washed out.

The first man to run Soap successfully was Clyde Eddy in 1927. But he had no idea what he had done. His geography was so confused that he had ordered his crew to portage at Badger, thinking it was Soap, then ran Soap, thinking it was Badger, at a flow of 40,000 cfs, when Soap almost ceases to be a rapid.

Now, still in the tail waves of Badger, I pivoted and rowed across the eddy fence—the interface of whirlpools between the downstream current and the reverse current of the eddy rushing upstream—and then on to the beach. The windy blast from an open oven seconds ago now seemed to come from a refrigerator.

The camp here on the Jackass Canyon side of Badger was a long beach stretching upstream to border the rapid. Scattered in the sand were huge slabs of tan sandstone that had peeled off the cliff thousands of years ago and now looked like the raw material for a do-it-yourself Stonehenge. As I pulled in, clouds of hot sand lifted off the beach and flew back toward Wyoming like horizontal rain, only to be deflected back toward the river by a seven-hundred-foot cliff. At its base, in more fallen slabs of Coconino Sandstone sand-blasted to silky smoothness, were some beautiful fossil footprints of predinosaurian reptiles.

I stared at the cliff. I had learned a lesson here, but not about footprints. On second thought, footsteps had been important. It was in September. I had just returned to the river after two weeks off—so my "sun clock" for Badger was a month behind. Michael Boyle warned me that I did not have enough daylight to hike everybody out of Jackass Canyon to the plateau, walk them the extra mile to the cliff overlooking this camp, and get them all back down before dark. But because I loved this hike and had just done it successfully a month earlier, I wouldn't listen. Boyle had been right, though; I couldn't pull it off this late in the year.

I had ignored other signs as well. When I gathered everyone for the hike, our three stewardesses asked me to wait while they finished putting on their makeup. Yes, stewardesses—one English, one Irish, and one Greek—for Saudi Arabian Airlines. The kind that we hear about from other trips but never see. But these were the real thing. All were in their twenties, and each had a perfect body. I would soon find out how perfect. They did not need makeup. They were runners. Maybe they ran to escape lusting Arabs.

We also had a seventy-one-year-old German I was worried about, but again I was wrong. At least I was consistent. Herman hiked like Reinhold Messner.

Another sign I ignored also came early, among the garage-sized boulders strewn in the Hermit Shale in the first third of the hike, where a couple who wouldn't watch those ahead of them never knew where to go next. They slowed us down, but still I was too stubborn to cut back on my plans.

When I finally got all of them up the rope at the twenty-five-foot waterfall and then on to the assault through the U-shaped canyon in the Coconino, Toroweap, and Kaibab formations, I still thought I had it made.

I made my big mistake right after we zigzagged to the top of the Kaibab and escaped Marble Canyon. Sunlight still burned the Echo Cliffs crimson, but only an hour of light remained. I should have congratulated them and then done an immediate about-face. Instead I tried to give them everything. I warned them that we were running out of light and *should* head back, but I said that if they hustled the mile out to the rim and back, they would get the view of their lives. Of course they all went for the view.

But as they marveled at the fantastic view down to Badger Rapid and our camp, I looked back at the Echo Cliffs, and my stomach sank. Direct sunlight had vanished. I rounded up everybody and hurried them back.

We were missing four people. Skip Horner was my only other guide, and this was his first time up here. I asked him to keep everybody moving while I ran back. Mike Anderson, our trainee, said he had tried to get the stewardesses and a woman from New York to leave, but they had shooed him away to take photographs.

When I got within fifty yards of the drop-off, the Greek with the classic body—who was wearing nothing but her shoes and a camera—motioned me to stop. The other three were naked and posing inches from that seven-hundred-foot drop. I regretted not having my own camera. Among other things.

When they saw I was serious about heading back immediately, they reluctantly put on some of their clothes. But when I explained that we were facing at least part of the descent down Jackass Canyon in the dark, they studied the alpenglow on the

50

cliffs and then gave me looks that said I was being a macho jerk. This irritated me. I told them to get dressed and run back to the descent into Jackass or they would hold up everybody.

The women of the friendly skies surprised me by throwing their clothes on and jogging back. But not the New Yorker. She would not even hurry. Even though we were on a road that a four-year-old could have run, she complained that she might sprain her ankle on a rock. The best I could do was convince her to walk a bit faster.

When we caught up with the group gathered at the head of the canyon, dusk had already made everyone a believer. Their faces asked me how I could have let all of this happen. I asked myself the same thing. I needed a plan.

I asked Skip to hightail it down to the boats and hustle back with every flashlight he could find. Meanwhile, we crawled down, dropping half a mile through the Toroweap and the upper Co-conino to the top of the waterfall. By this time it was nearly dark. I sent Mark down next to the pool at the base of the fall to spot people from the bottom for the last tricky move. One by one, they descended, gripping the rope as if a thousand-foot drop yawned below them. Old Herman climbed down like a mountain goat. Two really slow people gobbled all the remaining light by inching down in slow motion. The New York Princess was so worried that the only way I could get her to descend at all was to go with her. Of course, only one of us could use the rope, and of course it was her. While exposing myself to a pair of broken legs by jam-ming my fingers into the cliff and gripping her in case she peeled off, I had alternately sweet-talked her and rough-talked her down the fall. By then it was dark. Real dark. Dark thirty.

We gathered in the little amphitheater below. Every constel-lation in that narrow slit of sky was crystal clear. Moonrise was not due until just before dawn. People cracked jokes, but they probably had forgotten the potentially lethal exposures in the mile below us. They had never looked back on the way up to see them. Somebody asked me if all our hikes would be like this.

Skip arrived with five lights for the seventeen of us. Two were almost dead. Maybe this was just as well; people could see the route in front of their noses but could not aim the beams down the drops and scare the hell out of themselves.

But a funny thing happened. Sharing the flashlights brought everyone together. Within a hundred yards, a stewardess latched onto Herman. Sometimes one held *each* of his hands. The old goat could have out-climbed any of them, but he recognized opportunity knocking. He periodically went helpless to appeal to their maternal instincts. When we got back to Flagstaff two weeks later, he thanked me for the trip, confiding that it was the greatest trip of his life, "especially," he added with a wink, "the *first* hike."

Anyway, we made it back to the fossil footprints and the beach in pitch blackness in less than an hour—amazingly, without a scratch. Not even a prickly pear spine. It was incredible luck, and a cheap lesson.

Still grateful for that smooth escape, I stared at Jackass Canyon but thought now about those fossilized footprints of reptiles. Would a few of our own footprints fossilize here in the Canyon and someday be discovered by explorers of some strange race of intelligent life curious about the primitive creatures who preceded them? What would they guess that we were doing down here?

What *was* I doing down here? Instead of here, I belonged in an African rain forest. Reading *Tarzan of the Apes* as a teenager had hooked me on Africa's great apes—so much so that I had studied biology under the GI Bill, spent a year with gorillas, then two more with wild chimpanzees in a Ugandan rain forest to earn a Ph.D. in biological ecology. While at the University of California, I also learned how to row in white water as a commercial guide, which became my research support. I learned that male chimps wage territorial war against males of other communities for the possession of females—as men war against other tribes—

and I found out why. But by the time I graduated in 1980, seven years after the United States threw in the towel in Vietnam and ended the need for student deferments against the draft, no one was hiring Ph.D.s specializing in great apes.

Despite this sociological flip-flop, though, my stomach still growled when I got hungry. So I returned to rivers—here and in Ethiopia, Tanzania, Turkey, Sumatra, Java, Papua New Guinea, and Peru. I worked my way back to Kibale Forest in Uganda— only to have to leave again due to funding problems. I trekked among mountain gorillas in Rwanda and wild orangutans in Sumatra. Between these expeditions I always returned to the Grand Canyon. I could not help it.

All this was not just for adventure. I analyzed social behavior and concluded that many of the murky and illogical interpretations published by the experts and treated as gospel by many professors were wrong. But even after becoming an expert in the evolution of social behavior, meeting and marrying my wife, Connie, creating two little replacements for ourselves, Cliff and Crystal, and starting to teach in a university, I still found it impossible to stay away from Grand Canyon.

4

Shinumo

The rock here was slick enough to give climbers the heebie-jeebies. It was Redwall Limestone polished so smooth in places that some lizards have a hard time hanging onto it, and I was about to descend it wearing flip-flops.

These days we use a rope for the descent into Silver Grotto, the portion of Shinumo Wash (Mile 29) nearest the river—if we visit Silver Grotto at all. But now I was staring ropeless into the deep, bowllike concavity of the grotto, a fifteen-foot drop onto rock as hard and unforgiving as the Internal Revenue Service. This "route" without a rope makes only a few people think twice. Most simply do a careful about-face and inch their way along the narrow ledge back out to the bench and common sense.

A moment ago I had climbed the fifteen feet up from the boulder pile scoured by the river to the route into Silver Grotto. The cliff out there, away from the path of flash floods roaring down Shinumo Wash and above the high-water mark of the Colorado, was weathered and broken into convenient holds for hands and feet. It was designed to be climbed. But here . . .

Here it looks impossible. But with the right combination of friction points I could inch downward nearly halfway before being forced to run down the rest. A slip could mean waking up at the bottom with a broken leg, or worse.

I inched down, jamming the sides of my flip-flops against tiny irregularities in the rock only a centimeter wide. These had the uneasy feel of being too small for the job. They whispered, "Go back . . . " To grope farther down the wall, I suspended myself by fingers dug into too-shallow fingerholds to stretch and probe with my feet for more of those too-small irregularities. It was a question of balance and control, not necessarily mental. So, if this was crazy, why was I climbing down here?

It's hard to explain, but one look into the polished vault of Silver Grotto and the narrow green mirror of a pool nestled between its sheer walls reflecting their steep climb hundreds of feet up toward the broken thread of sky above is bewitching. Beyond this first emerald pool is a dry waterfall, then another pool between even more sheer walls, then another waterfall and pool, and so on ad infinitum it seems, and all concealed from the outside world. The only way to see any of this, even a small piece of it, is to climb into the grotto, then ascend by swimming the pools and climbing the waterfalls.

About halfway down the entry wall I had to shift gears. With no more finger- or toeholds, from here it would be a controlled run down the steep rock. The trick was not to slide at all, because once a slide begins, it's all downhill in an uncontrolled hurry where braking is accomplished by whatever part of the anatomy hits rock first. I let go and hit the sloping rock running.

Slap-slap-slap-slap, my flip-flops triple-timed down the polished limestone. Now to stop. Slap-slap-slap-slap, my flip-flops ran partway up the opposite side. I had stopped. And my feet were still in my flip-flops. I glanced back up at the face I had just descended. Good thing, I mused, that I did not have to climb back up that thing to get out of here.

I studied the pool that I had barely missed in my descent. "We swim from here," I said out loud. This pool was the only obvious route farther into Silver Grotto. While I knew I could circumambulate the pool by traversing the right wall to the fall—fun and good practice—it takes a lot longer. I dipped a toe into the mirrorlike surface.

Reluctantly, I slid entirely into it. As I had dreaded, the water was cold—warmer than the Colorado but still not warm. Then I started swimming. I detoured around the floating corpse of a bat, then stopped upstream at the inflow, now dry, where past waterfalls had carved a smooth, sinuous flume.

I wedged my back against the steep left wall and my feet against the right and inched upward using a chimney technique. Then I butterflied the flume with a hand and foot on either side, and lunged for the lip of the next pool. Easy.

The second pool, thirty feet across and a lot deeper than the first, again appeared the only conceivable route up the canyon, and this time it was. These walls are impossible to traverse without hammering hardware into the limestone. Before dropping into this bottomless *tinaja* (bedrock basin), I paused to study the polished organic curves and uncompromisingly vertical height of the silver walls rising in this ultranarrow slice in the Redwall. Silver Grotto itself appears an impossible place. Too narrow, too deep, too smooth, and with too much water in its bedrock pools. After all, this is the desert. Only about four inches of rain fall in this part of the Canyon in a normal year. But because the sun rarely hits down here, these deep pools evaporate very slowly. The water here is runoff from rains at least a month past—probably several months—and most of this water would remain here until the next rain, no matter how distant in the future.

Another few hundred feet ahead, Silver Grotto opened a bit wider into an amphitheater where bright sunlight penetrated to the canyon floor. The water in this pool was cold, but the sunlight ahead would be an antidote. I jumped into it, submerged, surfaced, then paddled toward the sun, driven like some primordial amphibian seeking the warmer ooze vital for its survival.

But the sunlight was attainable only after climbing another

flume in the bedrock, traversing yet another slippery pool, and climbing a very slippery cliff face of smooth gray Redwall. Finally I walked into the tiny, sun-warmed amphitheater near the upper end of Silver Grotto. This is still so deep in the limestone that from the top of the Redwall I would have appeared as big as a BB in a bucket.

Before me, in the center of the amphitheater, was another pool, this one steep-sided and the deepest yet, possibly twenty feet. Despite its being perfectly centered in the polished floor of creamy smooth rock and being symmetrical and glowing a luminous deep green that made it appear organic rather than geological, I felt no urge to enter it, nor did I have to. Along either side of this pool there was plenty of room to walk—until I reached the upstream end. There, as usual, was a steep, dry waterfall, this one about ten feet high. What lay above was hidden from view.

I walked around the left side of the pool to traverse up the waterfall. With shoes one can get a running start and sprint up this route with little reconnaissance. The hitch was that a failure during a running traverse left a lot of bruises. In contrast, a failure during a controlled traverse leads only to another swim. If you feel yourself slipping beyond the point of no return, you can kick off the wall to fall well out into the pool to clear the ledge hidden a foot or so deep in it and plunge safely into deep water.

I worked my way across the steepening limestone. Handholds became fingerholds, then became so tiny they almost vanished altogether at the final ascent. At that point—only seven feet below the lip of the falls in a very narrow section of canyon that debouched immediately into the pool and the sunny amphitheater—the only way to get up was to maximize one's friction against irregularities in the rock face, then "mantel" in a controlled pull-up/push-up. It is not a difficult technical move in bouldering, but neither is it a breeze.

I stretched upward full length and found the one critical fingerhold. This is the point of commitment. You must pull yourself over an abrupt lip of rock smoother than a baby's behind. The only way to do this is to control your friction and lift during the final lunge by maintaining three-point contact as you pull your entire body high enough to swing your right leg up onto the lip with enough body momentum to keep moving forward. Any slip, or too little momentum, and you go swimming. I made my lunge and then stood in the narrow opening to the next amphitheater. It was worth it.

This cavernlike room in the Redwall is perhaps the most beautiful in Silver Grotto. But this final traverse stops about half the people who visit here, even when they are offered assistance from above. This room is also as far as any human can go without a rope, a grapple, and a good throwing arm.

Before me lay a broad, wall-to-wall pool. This one was shallow at my end and green, with an underwater jungle of algae through which pollywogs fled like time-lapse elephants in a rain forest. A couple of water striders darted across the surface, their delicate feet almost dry because of their impossible ability to avoid breaking the water's surface tension. Nearby floated the corpse of a giant hairy scorpion, a victim of a fall from the unforgiving desert world beyond these walls. The pool spread like a sheet of emerald to the next dry waterfall, a sheer vertical drop of about thirty feet along rock so polished that not even a lizard could climb it. The wall to our right was undercut deeply and was luminous in sections where sunlight reflected shimmering from the pool. It was a place where humans were giants who added nothing but clutter but where some of the sparse but tenacious life of the desert had gained a tenuous and no doubt temporary foothold. I once witnessed this canyon funneling heavy rain and disgorging enough muddy river to float a small boat (though it would have been a suicide mission due to the huge falls). No doubt a major flash flood would again scour it to sterility someday. The desert offers none of the security of the earth's more humid life zones.

This room, this grotto, and this entire sinuous, smooth, deep canyon are all products of Redwall Limestone, the dominant cliff-forming rock in Grand Canyon. In fact, because it forms a cliff about five hundred feet high nearly everywhere, Redwall is the primary barrier blocking most potential routes into or out of Grand Canyon. Only where geological faulting has broken the cliff and allowed erosion to etch hand- and footholds is it possible to penetrate through the Redwall to escape or enter the Canyon.

Redwall is a marine limestone about 340 million years old, from the Mississippian age, an era when the first amphibians were crawling from the sea onto the land to start the biological revolution that led to the dinosaurs, to early mammals, and to river runners. Its true creamy gray color is evident everywhere here, but it contrasts with its normal red veneer of iron oxides washed down onto it from the Hermit and Supai formations above. Those rocks give the Redwall Limestone its seemingly contradictory name.

Far above my head, in fact, surrounding this elegantly carved grotto and extending far above it, is a geological textbook. Its pages are cliffs of stratified tans and reds climbing nearly 2,000 feet above the Redwall to the veneer of Kaibab Limestone. Although Marble Canyon here is barely half a mile wide—migrating bald eagles soar across it in seconds—these cliffs constitute part of an important chapter in Mother Nature's logbook—a book that erosion seems determined to condense in a way that would make even the people at *Reader's Digest* blush.

These strata escaped the massive sheet erosion caused by the uplifting of the Colorado Plateau. During the past 20 million years or so, erosion stripped many of the other, newer pages of that logbook. Entire landscapes of strata a mile thick atop the Colorado Plateau and dating from the time of the earliest dinosaurs 225 million years ago to the rise of mammals after the Cretaceous extinctions 65 million years ago have vanished from this region. Written in the rock here and in the Canyon ahead, though, is ten times more planetary history than that 160-million-year reign of the dinosaurs.

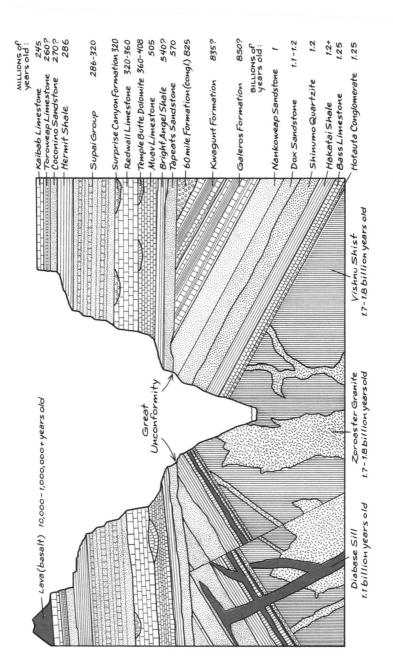

MILLIONS of years old:

Kaibab Limestone 245
Toroweap Limestone 260?
Coconino Sandstone 270?
Hermit Shale 286

Supai Group 286-320

Surprise Canyon Formation 320
Redwall Limestone 320-360
Temple Butte Dolomite 360-408
Muav Limestone 505
Bright Angel Shale 540?
Tapeats Sandstone 570

60 mile Formation (congl) 825

Kwagunt Formation 835?

Galeros Formation 850?

BILLIONS of years old:

Nankoweap Sandstone 1
Dox Sandstone 1.1-1.2
Shinumo Quartzite 1.2
Hakatai Shale 1.2+
Bass Limestone 1.25

Hotauta Conglomerate 1.25

Lava (basalt) 10,000 - 1,000,000+ years old

Great Unconformity

Vishnu Shist 1.7-1.8 billion years old

Diabase Sill 1.1 billion years old

Zoroaster Granite 1.7-1.8 billion years old

FIGURE 4.1 The Geology of Grand Canyon (Sources: Breed and Roat, *Geology of the Grand Canyon*; Stevens, *The Colorado River in Grand Canyon: A Guide*; Terry Wright, Wilderness Interpretation [Forestville, Calif.])

The rimrock of the Canyon, the Kaibab Limestone, is about 230 million years old. Surrounding me now at river level, the Redwall Limestone is 110 million years older. Between the two are other sedimentary formations: the Toroweap Limestone, Coconino Sandstone, Hermit Shale, and the Supai Group. During the 110 million years when these rocks were building up, this region straddled the equator thousands of miles from here. Instead of being a mile and a half above sea level as today, the Colorado Plateau was a broad continental shelf of the supercontinent of Pangaea that the seas repeatedly transgressed to cover and regressed to expose—as if it were being washed by slow-motion tsunamis millions of years apart.

When the transgressing sea covered it, the rock built up. But when the sea receded, exposing the rock to rain and wind, erosion scoured it to create a clean, new interface on the landscape—what geologists call an "unconformity," which can represent the loss of hundreds of millions or even billions of years of deposits a mile or more thick. In this cliff every chapter of earth history that seemed to end as if a cosmic floor grinder had polished it was an erosional unconformity.

It used to bother me how different formations had formed atop one another here in the "same" place. Why wasn't new sandstone deposited atop old sandstone, and shale atop shale? I guessed it was because, over time, mountains of different rock had eroded and washed down to the sea. But this was wrong. Different sediments were always carried to the sea. Different rocks were deposited atop one another in the same place—like a buckwheat cake on a crepe on a buttermilk pancake—because the ocean was "nomadic."

It works like this. When a river empties into the ocean, its flow decreases as it joins an oceanic current flowing parallel to the coast. The first sediment to drop out, the sand, is the heaviest. Hence, beaches are located along the seashore, not a hundred miles out in the ocean. This also explains why sandstones originate near shore. Farther out to sea, as the river slows more, ever smaller particles, like clay, drop out to create mudstones and shales on the seafloor in broad bands that parallel the shore.

When, after millions of years, the sea transgresses and causes the coastline to creep inland, the place where sand used to settle will be farther out in the ocean and will instead receive silt. Thus a shale will form atop the sandstone. Meanwhile, new sandstone may literally follow the seas inland and continue to form on what was once dry land—atop the unconformity created there by erosion. When the sea rises farther, the shale atop the older sandstone may itself be buried by a limestone typical of quiet ocean. If you sliced through all this, as the river has done here, you'd find all three rocks, one atop the other, but all in the same "place," exactly what we see here.

This cliff is Redwall, a marine limestone deposited in quiet ocean partly as a buildup of the calcareous exoskeletons of sea creatures. The formation above it, also about 500 feet thick, is the Supai Group of sandstones and shales deposited much closer to shore, where coastal currents slowed. Above the Supai is the thick red Hermit Shale, deposited an intermediate distance from shore and sprinkled with tropical ferns washed down from primeval forests. Then, after a receding of the sea here, Coconino Sandstone was deposited offshore, then reworked by wind onshore as sand dunes, now petrified, typical of a Saharan desert but crawling with the reptilian forebears of the dinosaurian revolution. Next, millions of years later, after the Coconino desert submerged under a rising sea, thin beds of Toroweap sandstones were deposited on top of it, then Toroweap shales and mudstones and limestones, then more Toroweap shales again as the sea transgressed and then regressed again, as if in indecision. Finally the sea rose a great distance to cover the whole region under quiet ocean and bury the landscape under Kaibab Limestone. All of this happened during a 110-million-year tectonic ride of the future North American landmass immediately prior to its slamming into Africa to become part of the supercontinent Pangaea.

Although it may sound crazy, particle astrophysics makes all of this inevitable. This planet—and our solar system—was born in the aftermath of a gigantic supernova, a mammoth atomic explosion 6 to 10 billion years ago that sent trillions of trillions of photons and other primordial subatomic particles streaming into infinity and also created a local cloud here. A few billion years later, gravity within this cloud formed a protosun that started spinning due to other gravitational currents in interstellar space. Its spin marshaled the cloud of gas and dust surrounding it into an accretion disc similar to but immensely more massive and hot than those now orbiting Saturn, Uranus, and Jupiter.

This solar disk condensed further into particles, then ever larger accretion bodies—including planetesimals and planets. Temperatures were so hot within the inner solar system that accretion there included more high-temperature minerals, which led to the formation of the terrestrial planets—Mercury, Venus, Earth, Mars, and our moon—roughly 4.6 billion years ago. Due to the colder temperatures far beyond the orbit of Mars, the outer planets became gas giants.

Earth is just one in Sol's cosmic litter of nine planets, but it differs even from the other four terrestrial planets. This is because Earth's 100 million years of accretion released so much gravitational energy and such tremendous heat from cosmic collisions that it melted many minerals. Once the temperature soared to the melting point of iron, it triggered the Iron Catastrophe: a sudden internal meltdown that plunged heavier minerals toward the core, thus releasing even more gravitational energy and heat and causing more melting, and so on. This is why minerals near the surface are predominantly the lightest, with water on the surface and gasses trapped by gravity above it.

Unfortunately, this planet destroyed most of its own early history. Knowing anything at all of earth history—let alone of Grand Canyon—depends on dating the rock. Rock is either igneous, sedimentary, or metamorphic. Igneous rock originates from a molten flow. Sedimentary rock derives from particles deposited either chemically or by wind or water and then cemented together. Metamorphic rock is either of the first two that has been altered in structure by heat or pressure.

Sedimentary rock is often dated by looking at its physical properties and by correlating its fossils—especially those that became extinct at a known time—against those in other formations. But this yields only a relative age. Paleomagnetism is the best bet to gain an absolute date for metamorphic rock. In Grand Canyon the metamorphic Vishnu Schist fifty miles downstream from here in First Granite Gorge was dated via paleomagnetic orientation to about 1.8 billion years ago. With igneous rocks that seal and trap their own argon from the radioactive decay of the isotope potassium 40—at a half-life of about one and a quarter billion years—radiometric dating yields an exact age.

Even with these techniques, though, until 1989 it remained a mystery what continents existed two billion years ago. The solution to this mystery explains why rock in the lowest depths of Grand Canyon came to be as it is.

North America is composed of seven microcontinents, or continental plates. Laurentia, the name for the ancestral North American supercontinent, was assembled when these first seven, already old themselves, slammed together between 1.96 and 1.81 billion years ago—a surprisingly short time for so many collisions. What is important to understanding Grand Canyon is that this assemblage (geologist Paul Hoffman refers to it as the United Plates of America) left one of its trademarks fifty miles downstream from here: extreme metamorphosis.

The microcontinent containing the American Southwest collided and bonded with Laurentia roughly 1.8 billion years ago. Next, the microcontinent containing California, Nevada, and the landmass extending north to Alaska slammed into Laurentia and the region containing the future Grand Canyon. Along the interface between these two colliding plates, a giant mountain range thrust upward. The important point is that this collision and mountain building almost certainly caused the severe metamorphosis of the Vishnu Schist that made up that mountain range. Possibly a billion years passed before plate tectonics nibbled northern Europe and Greenland from this Laurentian supercontinent and set them on their independent destinies.

The landscape here has been strangely quiet since about 570 million years ago despite being lifted a couple of miles above sea level. This stability is also surprising in view of the thousands of miles that North America has traveled in those half a billion years and in view of its collision with South America and Africa to form the even more gigantic supercontinent of Pangaea 300 million years ago. Of all great plateaus, the Colorado—the world's second highest great plateau—is unique due to its having risen gently 7,000 to 13,000 feet while its strata remained undisturbed.

The "father" of the theory of continental drift, Alfred Wegener, died in a blizzard during a field trip in Greenland before his theory was vindicated, but today his insights are the key to understanding the shape of the world. They are as important to understanding geology as Charles Darwin's theory of natural selection is to understanding biology. During the 1950s, paleomagnetic data revealed that the opposing coasts of continents shared identical geology but were separated by thousands of miles of very different rock under the seas. Echo sonar mapping of the ocean floor during the 1960s uncovered extensive midoceanic ridges 46,000 miles long in the Atlantic and Pacific. New seafloor was emerging from the crust here at these midoceanic ridges, spreading outward toward the continents, colliding with them, and apparently shoving them apart. Because the seafloor is composed of basalts that are denser than the continents, the colliding seafloor sometimes subducted under their edges. In other cases, such as along the east coast of North America—the ancient coast of the Colorado Plateau—the continents were merely passively shoved. These were the first big clues to the mechanism of continental drift: plate tectonics.

They also explained why the oldest seafloors were a mere 200 million years old: this was the longest it took for them to travel between a midoceanic ridge and the edge of a continent. Further, despite the fact that subducting plates move less than an inch a year, 14 cubic miles of lithosphere (crust) dives underground into the lower mantle annually along subduction zones, producing a violent ring of earthquake faults and active volcanoes. (The 30,000-mile periphery of the continental plates surrounding the Pacific Ocean, for instance, is dotted with 75 percent of all the active land volcanoes. Geologists call it the Ring of Fire.) Again, Grand Canyon would not exist but for these geodynamic processes.

Recently a series of satellites of the Global Positioning System orbiting 12,000 miles above the earth not only revealed that the continents are indeed moving, they also tell us their precise progress. What they don't tell us is that the driving force behind plate tectonics is heat deep within the Earth generated by the decay of isotopes of strontium, rubidium, samarium, neodymium, uranium, lead, and several other elements. This decay has been going on since the beginning of this world, and it is what keeps the core extremely hot—possibly 5,600°—as hot as the surface of the Sun. This heat is ultimately what slams continents into one another and raises huge plateaus like Tibet and the one through which the Colorado River sliced Grand Canyon.

The mechanics of this are simple even though Earth is the only planet in the solar system known to experience plate tectonics, which apparently results from its position in the solar accretion disk. By monitoring the propagation of shock waves transmitted through the Earth during earthquakes via seismic tomography, geologists have identified gigantic hot blobs thousands of miles across, rising about an inch a year in the Earth's mantle. These swell until they reach the lithosphere, where they move laterally, cool somewhat, then sink. This convection drives continental drift. Interfaces between two rising cells in the mantle create midoceanic ridges and cause them to extrude new seafloor. Before cooling enough to sink, the blobs move horizontally along the base of the crust, dragging the new seafloor crust (three to five miles thick) toward the continents. They also drag or shove along the deep-rooted (up to 250 miles thick) continental crust, thereby causing continental drift. In effect, the continents are ancient rafts of lighter rock that have floated on the upper mantle for billions of years.

But most important to understanding what one is really seeing in the fantastic cliffs here in Grand Canyon is how these blobs explain the advance and retreat of the seas, seesawing back and forth to create a spectacular sedimentary "pancake stack" of geology.

When a rising blob hits oceanic crust, geologist Brad Hager points out, the crust probably bulges upward and makes the sea shallower (as T. A. Heppenheimer reported in 1987). The "extra" water rises and transgresses inland. Sedimentary rock is then deposited on a region of formerly dry land. When the blob cools and descends into the mantle, the sea level lowers and regresses from the continent. The surface of that new sedimentary rock then erodes, and it will be separated from the next depositional event by an unconformity. This fits exactly with the scores of such events visible in the sedimentary walls of Grand Canyon.

But convection does even more; it assembles supercontinents. In computer simulations, microcontinents are drawn to regions where large and consistent down-welling of blobs occurs such that eventually the plates all collide with one another and bond into supercontinents. Hence the surprisingly rapid assembly of Laurentia, the North American supercontinent.

But supercontinents such as Laurentia and Pangaea may be suicidal. Seismologist Don L. Anderson proposes that, once assembled, supercontinents form a "thermal cap" atop the mantle that, because of the supercontinents' great thickness, reduces heat loss from below and causes such a rise in mantle temperature that it alters the convection patterns. Superswells (or superblobs) of mantle material then rise directly under the continents and blow them apart.

Two strong lines of evidence suggest that this actually happens. First, supercontinents do appear unstable. They always seem to break up as if forces in their centers were shoving them apart. Second, a common event a hundred million years or so after the assembly of supercontinents is the massive intrusion of igneous red granites into the continental crust and ryolites onto its surface. Red granites are so pronounced 1 to 2 kilometers deep in the center of North America that geologist Marion Bickford points out that in "essentially every drill hole that goes down, that's what you get." Again, the evidence is obvious here. Everywhere in Grand Canyon the 1.8-billion-year-old Vishnu Schist is riddled with a red granite called Zoroaster Granite. Red granites seem unconnected with mountain building; instead they appear within the unstressed crust of large continents exactly as if superswells in the mantle were melting the granites below and forcing them upward.

I stared again at the emerald pool bracketed by the first impassable cliff-waterfall. Several grottos, amphitheaters, pools, and steeper falls exist above this last one in Silver Grotto proper. I have climbed into them but not from here. A mile downstream from Shinumo Wash, faulting has rubbled the Redwall to open a route to the top. After traversing to the top of the cliff and then hiking inland to the head of this canyon, one can descend Shinumo into the upper section of what might still be called Silver Grotto. But I did not drop to immediately above this waterfall. A couple of ridiculous falls stopped me. I had not thought to bring a rope. Without one, my descent would have become an interesting way to die—like an ant trapped in a goldfish bowl.

I now retraced my steps to the top of the waterfall at the head of the sunny amphitheater, then descended. The amphitheater was peaceful—almost timelessly so—but again this peace was deceptive; flash floods scour this wash periodically and violently.

I could not tear myself away yet. The walls were so smooth and symmetrical surrounding the deep oval pool that the place seemed a tectonic womb, an inorganic birthplace for geological phenomena as yet unwitnessed and undreamed of by science. I stared at the silent pool. I imagined something emerging from it, as if in spontaneous birth—maybe one of those primordial amphibians arriving eons too late to enter the evolutionary race toward sentient mammalian bipeds.

But the surface of the pool did not ripple. Somehow the reality of its silence made Silver Grotto even better. Suddenly I felt so lucky to be here that I laughed out loud at my own good fortune. Turning my back on the pool took an effort of will. Even so, once at the bottom I could not resist taking a last look into that narrow secret world. Yes, I was lucky. We all were. Silver Grotto and nearly all of Marble Canyon have been targeted by the hydropower men in Washington to be buried under a mammoth reservoir.

5

Entropy and the Sacred Cows

The creaking of a bowline stretched taut by the dropping river woke me. I pried my eyes open to gaze into the sky. Stars twinkled in the blackness. I squinted to focus. Taurus dominated the narrow slice of sky above. I looked at my watch. Four-thirty. Dawn would fade those stars in twenty minutes.

All five of our boats were strapped together, and with their bows now stranded on the beach due to the Bureau of Reclamation's choking the outflow from Glen Canyon Dam last night, they were stable and quiet. Soon they would tilt so much against the steep beach that I would start sliding off my rear deck into the bilge—soon, but not yet.

This was not my morning to cook breakfast. I dropped my head back onto my pillow, a Patagonia jacket bundled into a stuff sack secured to my ammo box by a carabiner, and stared at the mammoth cliffs of Redwall that have guarded the river for millions of years. Somehow the clean, sharp faces appeared much younger, as if carved from the bedrock during the past few millennia. I closed my eyes again, but the cliffs persisted in my mind. I gave up and opened my eyes. These cliffs were nothing to take for granted.

Here at Nautiloid Canyon (Mile 34) they appeared unreal in

the starlight—steep, sheer, and impossibly high. A fantasy landscape. But this inner canyon of Redwall was real, almost too real. The impervious look of the Redwall made this the kind of river canyon over which dam builders drool. And the engineers and politicians of the Bureau of Reclamation had done more than drool. They had sent a reconnaissance force to find the best location to implant a dam that would drown Marble Canyon all the way upstream past Lee's Ferry into the mouth of Glen Canyon. For those who loved Grand Canyon, these surveyors were the harbingers of Evil. After completing their survey, the bureau decided to build Marble Canyon Dam five miles downstream from here. For us who run this river—every boatman is a conservationist—this decision was premeditated murder.

Why dam Grand Canyon? To make money. But the bureau's financial equation was not a simple one: the money would not so much be made as transferred via machinations from the tax coffers of the federal government to the pockets of contractors building the dam, the brokers of power and water, and the seekers of the water rights who own or would buy the acres of basically worthless land that federally subsidized water from the Colorado would convert into prime real estate sellable at astronomical profits. If this seems a cynical view of what the bureau touts as progress, judge for yourself.

Although the entire Southwest has serious water problems, Arizona today is running on empty with the abandon of a cocaine addict in the fast lane. Half of Arizona receives less than ten inches of rain annually, much of it a lot less. Paradoxically, most of Arizona's four million residents live in Maricopa County's south central valley and the greater Phoenix metropolitan area—one of the least rainy regions in the state. More astonishing, Maricopa County is the fastest-growing area in the United States; by 1990 Phoenix had become its tenth largest city. How do they do it?

With groundwater. Maricopa County annually pumps two million acre-feet of groundwater more than what nature replenishes from an aquifer charged during the Pleistocene (an acre-foot is 325,851 gallons). For reference, 13.9 million acre-feet is the average annual flow of the entire Colorado River. Although agriculture accounts for 90 percent of Arizona's total water consumption (and cattle feed accounts for most of that), Phoenix residents use 50 percent more than the national average of 36,500 gallons per capita. Fountains and artificial lakes built to sell land developments dot a landscape of lunar aridity where summer temperatures below 100° are considered cool. Phoenix also holds the title as the Air Conditioning Capital of the World. Again, the river that was shrinking beneath my stern generates a lot of the electricity used to chill that desert city. It even fills many of those fountains and artificial lakes.

The Bureau of Reclamation's Salt River Project traps most of Arizona's annual surface runoff—a mere 2 million acre-feet—before it can join the Colorado. Consequently, 80 percent of the water used in the state is pumped from aquifers. When Phoenix was first settled, water flowed from artesian springs. Since 1909 at least 130 million acre-feet have been pumped from the south central Arizona aquifer (equivalent to ten years' flow of the Colorado). Maricopa County is mining its groundwater so heavily that some regions are losing elevation. Some wells drilled thousands of feet deep to chase the sinking aquifer hit only hot brine. Today the Salt and Gila river system and Maricopa County's improvident use of groundwater barely sustain demand. In the near future the vanishing aquifer will herald a disaster.

The problem was exacerbated by Arizona's having fallen behind Los Angeles in the stampede to claim the Colorado. In 1922 anxiety among the Colorado basin states over the loss of water to rapacious Southern California prompted Herbert Hoover, then secretary of commerce, to travel to Santa Fe to preside over the formation of the Colorado River Compact. Eleven months later the state delegates agreed to divide the river equally between the upper basin states of Wyoming, Utah, Colorado, and New Mexico and the lower basin states of Nevada, Arizona, and California.

72

They arbitrarily drew the line through Lee's Ferry to divide the "basins"—this decision resulted in Lee's Ferry becoming Mile 0 for all distances measured downstream. (Worried about the portion promised to California, Arizona refused to ratify the compact until 1944.) Then, relying on twenty-five years of flow records that included the wettest ten-year period on record and that were known to be unreliable, the Reclamation Service estimated the annual flow of the Colorado at Lee's Ferry to be 16.8 million acre-feet. Improved measurements have since revealed that the river has delivered only 13.9 million acre-feet of water annually during this century. The bottom line? Those faulty data resulted in a compact that annually promised the lower basin 3 million acre-feet of river water that did not exist. That promise still stands.

Today the upper basin states are obligated to deliver 8.23 million acre-feet to the lower basin yearly. Of course, this obligation itself cannot conjure the missing trillion gallons per year that the river does not have. But because the upper basin states and Arizona lacked both the delivery systems and a population with a need for their allotments, the paradox of nonexistent water being guaranteed by law led to no immediate crunch. Even so, the early recognition that the compact was incompatible with nature did lead to two solutions: first, the bureaucrats branded the Colorado a "deficit" river, as if it had failed in its job; second, they decided to build a huge dam to store high flows from Grand Canyon, a dam that would convert 14 percent of the Grand Canyon of the Colorado into a reservoir.

A decade after the 1922 compact became law, the Bureau of Reclamation (and, in lower profile, the Los Angeles and Coastal Metropolitan Water District) accepted the low bid of $48,890,000 from a consortium of eight fledgling companies, including Bechtel and Kaiser, to build a 726-foot dam in Boulder Canyon. In 1935, after two years of pouring 3.4 million cubic yards of concrete and less than three years of construction, the Colorado River experienced its first no-holds-barred assault by humanity: Hoover Dam, heralded as the single greatest engi-neering achievement of the human race. It was the largest thing humans had built in the history of the world—6.6 million tons of concrete (falsely reputed to entomb a few corpses of the 110 workers who died during construction) blocking a river that ranked twenty-sixth in size in the United States. This most gi-gantic of monuments was constructed when the population of the West, about 11 million, equaled that of New York State. Rather than having been built for a pharaoh or some megalomaniacal emperor, the epic monument of Hoover Dam was constructed for sacred cows.

The bureau was quick to point out that the 28 million acre-feet that Hoover Dam impounded in Lake Mead, creeping forty miles into Grand Canyon, were too few. More dams on the Col-orado were vital to getting the job done. What job? The bureau's job: capturing water to "reclaim" arid lands through irrigation.

In most cases, however, especially in the upper basin states, where soils were poor and the growing seasons short, the dams and irrigation works were so ridiculously expensive that no farmer, no matter how large, could afford to buy the bureau's water at its real cost. Their crops did not earn enough. Even with massive federal subsidies, these farmers could never repay the cost of building the bureau's projects. This incompatibility be-tween the bureau's "vision" and fiscal reality almost stopped the dam builders.

Ironically, the high dams themselves offered the solution. When Hoover Dam went on-line in 1936, it was the largest elec-trical generator in the world. In the 1940s, Congress sanctioned the bureau's proposed modus operandi of building dams strictly for the revenue they produced through the sale of hydroelectric power. By this time, as Russell Martin reports, bureau hydro-electricity had built 60,000 aircraft critical to winning World War II. Such dams became known as "cash registers." Via fiscal leg-erdemain, the revenues from cash-register dams were retained by the bureau to bankroll otherwise unjustifiable water projects whose cost-benefit ratios could not stand up even to the bureau's own distorted analyses.

Once the concept of cash-register dams invaded the West, its rivers became sacrificial victims. On the Colorado River, Hoover Dam was joined by (starting upstream) Fontenelle Dam, Flaming Gorge Dam (on the Green), Glen Canyon Dam, Parker Dam, Davis Dam, Headgate Rock Dam, Palo Verde Dam, Imperial Dam, and Laguna Dam, plus Morelos Dam in Mexico— plus many more dams on its tributaries: Blue Mesa, Morrow Point, Crystal, and Curecanti (in Colorado on the Gunnison and Colorado), McPhee (on the Dolores), Navajo (on the San Juan), Alamo, Theodore Roosevelt, Horse Mesa, Mormon Flat, Stewart Mountain, Horseshoe, Bartlett, Coolidge (these last eight, in Arizona, completely stopped the flow of the Verde-Salt-Gila system into the Colorado), plus Seedskadee, San Juan–Chama, and Paonia dams (in Utah). These dams store a total of 61.5 million acre-feet and can generate 3.624 million kilowatts of electricity. The Bureau of Reclamation transformed the Colorado from an impressive force of nature running rampant through the West to megalithic plumbing. In *Rivers of Empire* Donald Worster summed it up: "the Colorado had become . . . a part of nature that had died and been reborn as money."

But did bankrolling dubious water projects via cash-register dams work out as well in practice as on paper? No. In fact, the projects did not look good even on paper. After World War II, Senator Paul Douglas of Illinois reanalyzed the bureau's plans for the Colorado. Using faultless arithmetic, Douglas explained that the Colorado dams would generate kilowatts costing about 500 percent more than existing projects in the East. He also revealed that irrigation projects in the upper basin states, particularly the largest one proposed, the Central Utah Project, would cost "about $4,000 an acre—six times the cost of the most fertile land in the world." Douglas further pointed out that 95 percent of the acreage proposed in these projects was to be used for cattle. In other words (as Reisner quotes Douglas), the bureau would be making "an average expenditure of $2,000 an acre for land which, when the projects are finished, will sell for only $150 an acre." Nevertheless, in an act that is appalling to river runners today, Congress gave the bureau the green light for most of what they asked anyway.

Did the dams do the job? Financially, they were ludicrous. No entity in the world other than the federal government with its budget apportioned by a Congress ruled by pork-barrel deals could have backed them. But even worse, a close look produces the nagging conclusion that the bureau plugged the Colorado with too much of a "good" thing.

The average loss of Colorado River water due to evaporation from the bureau's large reservoirs adds up to 2.244 million acre-feet yearly, almost equaling (and therefore doubling) the annual "deficit" the Bureau of Reclamation finally imposed on the Colorado, 2.5 million acre-feet. Indeed, in the late 1960s the bureau admitted that an annual flow of 20.2 million acre-feet (145 percent of what the Colorado actually delivered) was necessary to justify constructing all 134 of their proposed projects. By now the bureau had to admit that their plans for the Colorado, particularly the pending Central Arizona Project (CAP), exceeded what the river held and would require an augmentation plan to import a minimum of 2.5 million acre-feet. This was just one bit of illogic in the ill-conceived Central Arizona Project.

In 1964 Arizona won a twelve-year suit against California and gained its 2.8 million acre-feet of the Colorado, a million of which went to Arizona Indians. This spurred action on the Central Arizona Project, which is a system of 333 miles of canals, 14 pumping stations, and holding reservoirs, all designed to pump at least a million acre-feet out of the lower Colorado at Lake Havasu and move it up and over several natural obstacles, lifting it 2,909 feet on a journey to a final destination 1,249 feet above the Colorado. The estimated cost to the American taxpayers was $3.5 billion for the canals, dams, and pumps. The energy required for continuous pumping of at least a million acre-feet of water half a mile above the river each year was a logistical nightmare. In his excellent book *Cadillac Desert,* Marc Reisner sums it up: "Using the most optimistic set of circumstances—high-value crops, unprecedented crop prices, dirt-cheap power from preexisting dams—the Central Arizona Project was likely to be an economic catastrophe."

But Arizona's veteran senator Carl Hayden was chairman of the Senate Appropriations Committee. He was also a man who had pushed hard for Arizona's share of the Colorado during his 58-year career in Congress, and he was the single greatest congressional ally of the most notorious commissioner the Bureau of Reclamation ever had, Floyd Dominy.

In 1964 the bureau (aka Floyd Dominy) unveiled its Pacific Southwest Water Plan, an audacious blueprint to repair the West's "deficit" river and to insure completion of the CAP. It first required two dams in Grand Canyon, one near Mile 39—five miles downstream from where I was watching the first glow of dawn at Nautiloid Canyon—and a second near Bridge Canyon, near Mile 238. Neither the proposed Marble Canyon Dam nor the Hualapai (Bridge Canyon) Dam would have dammed the river within what was then Grand Canyon National Park. When the park was established in 1919, Carl Hayden (already a member of Congress) was instrumental in whittling it down to 645,000 acres and in gerrymandering its boundaries to exclude from the park huge areas of mining, timber, and grazing. He also included a clause reserving space within it for dams "when consistent with the primary purpose of said park."

It was not until 1975 that President Gerald Ford signed the bill sponsored by Barry Goldwater and Morris Udall doubling the park to 1,892 square miles to encompass the entire 277 miles of the river in Grand Canyon, including the last 40 of those miles drowned by Lake Mead. But in 1964 the Colorado within the park began at Mile 52 (Nankoweap Creek) and ended at Mile 134 (Tapeats Creek). Instead of all 237 miles of river, the park contained only 82. The Marble Canyon and Hualapai dams would have drowned approximately 133 of those miles outside the park but inside Grand Canyon.

Why build them? Not to store water. The Colorado River no longer flows to the Gulf of California. Every last salty, recycled drop of the thoroughly used Colorado (one estimate reckons that each drop of the Colorado is used at least three times) sinks into an even saltier field in Mexico. On most days the mouth of the river is a brackish mud flat. Neither dam could catch and store extra water, because none exists. In fact, the building of either Marble Canyon or Hualapai dam would result in a net loss of Colorado River water due to extensive additional evaporation. So why build them? Because they were planned as cash-register dams to produce revenue by generating electricity. But revenue for what?

This is where the Pacific Southwest Water Plan goes beyond audacious to become insane. Part of the revenue from the proposed dams within Grand Canyon was to provide power for pumping the Colorado uphill to Phoenix and Tucson—to reverse entropy and make the CAP flow against gravity. That was the audacious part. The insane part was to use the revenues gained from brokering the remaining hydroelectric power to help finance a pair of huge dams on the Trinity River in northern California plus a long tunnel through the Trinity Alps to send Trinity River water to the Sacramento River, where it could then have a semiconventional engineering chance of being pumped (uphill again) to Los Angeles.

Why to Los Angeles if all these dams were intended for the Central Arizona Project? Because sending nearly 2.5 million acre-feet of the Trinity to Los Angeles would free an equal amount of the Colorado, which Los Angeles would no longer receive (want to bet?). The Colorado savings then would be sent to south central Arizona via the CAP. The costs of the Pacific Southwest Water Plan—several billion dollars plus drastic ecological destruction—were rationalized as reasonable sacrifices by the American people to satisfy the greed of private land developers and the dreams of ranchers with their sacred cows.

Why do I say *sacred* cows? By 1985, according to Tad Bartimus, the Bureau of Reclamation operated 333 reservoirs, 345 diversion dams, 990 miles of pipeline, 230 miles of tunnels, 188 pumping stations, 50 power plants, 14,590 miles of canals, and 35,160 miles of smaller laterals in seventeen western states. The bureau had spent 11 billion tax dollars on all this by 1984. In the upper basin states, 90 percent of the Colorado's water that is used goes to irrigation, and 85 percent of that water irrigates cattle feed. In the lower basin states, 85 percent of the Colorado is used for irrigation, and 82 percent of that is used for cattle. The river irrigates a total of 3.4 million acres. In the West it takes 4,200 gallons of water to produce a pound of beef (compared to 300 gallons to grow the wheat for a loaf of bread). The Colorado now feeds about 15 percent of America's 100 million cattle. (Roughly one-third of North America is devoted to grazing, and each pound of beef produced also costs us an average of about 35 pounds of eroded topsoil. In California, irrigated pasture consumes 4.2 million acre-feet, one seventh of the state's water, to add only one five-thousandth—$94 million—to the state's economy.) If all those figures are difficult to interpret, the bottom line is this: 74.4 percent of Colorado River water used in the United States—and sold at 10 percent of its development cost—goes directly into maintaining roughly 15 million cows, the most expensive cows on this planet. To spend that much money on them, they must be sacred to someone; clearly they are to the Bureau of Reclamation.

As a by-product, Donald Worster reports that in their efforts to "reclaim" America west of the hundredth meridian, the bureau "had forced out of use 5 to 18 million farm acres in the East . . . sending thousands of rural men and women into bankruptcy." Worse, farmers on well-watered lands east of the hundredth meridian were being paid *not* to grow nonlivestock crops that westerners do grow on desert lands irrigated at huge expense by the bureau so that westerners could grow those crops instead. Worster also gives Raymond Moley's mid-1950s assessment of the bureau's modus operandi: "Money was being taken from the American public in the form of taxes and redistributed according to the social values of powerful bureaucrats, and those bureaucrats favored western farmers over eastern farmers, over urban dwellers, and over industrialists who wanted water too." Grand Canyon was to be dammed twice to finish this massive makeover of the West.

Marble Canyon and Grand Canyon escaped being sacrificed to the bureau's "monument syndrome" due to the heroic efforts of one determined man. That man was David Brower, executive director of the Sierra Club. A decade earlier, Brower had fought to save Dinosaur National Monument from two bureau dams proposed for the Yampa and Green rivers. The cost of victory had been a tacit agreement to let the bureau dam little-known Glen Canyon instead. Afterward Brower floated the 170 miles of the Colorado in Glen Canyon. By the time he reached Lee's Ferry he realized he had made a serious mistake—in fact, he considers not fighting Glen Canyon Dam the single greatest failure in his life.

In an almost mythic way, the 1960s conflict between Floyd Dominy and David Brower over damming Grand Canyon became the classic struggle between the good of conservation and the evil of environmental destruction for what are known well in advance to be negative results, and although the battle to save Grand Canyon involved more people, institutions, and organizations than Brower and Dominy, they were as much the principal adversaries of this battle as David and Goliath were of theirs. Their conflict set new standards for weaponry used in environmental struggles, weaponry that pulled Grand Canyon from the jaws of a bureaucracy whose ethics had long been subverted by political brokerage.

David Brower had been a conservationist at heart ever since

he had spent his sixth birthday below Hetch Hetchy Valley as the infamous dam was being built (within Yosemite National Park) to flood Yosemite Valley's twin sister. Decorated four times in combat during World War II, Brower also held no illusions about what it took to win.

Floyd Dominy, in turn, had been a dam builder since the day in the late 1930s when he first sat on a bulldozer in Campbell County, Wyoming, and graded his first stock pond in a range desiccated by drought. Dominy considered dams both the ultimate ascension of man over cruel nature and a great way to make a living.

It was a grudge match of epic proportions. The prize was the fate of one of the Seven Natural Wonders of the World—and the free-flowing river in it.

Dominy shoved the Colorado River Basin Storage Bill (including part of the Pacific Southwest Water Plan) into Congress, at which point he considered the cash-register dams in Grand Canyon as good as built. Even the National Park Service had turned Judas. Philip Fradkin, in *A River No More,* reports Frederick Law Olmsted, Jr.'s evaluation in the Park Service's own study that a high dam in Marble Canyon "might not be objectionable from the scenic and recreational viewpoint." Olmsted also favored the high dam at Bridge Canyon. Fradkin further reports that Carl Hayden warned Secretary of the Interior Stewart Udall, "We have got to get the bird watchers in line, and, by the way, the very nature lover we want to get in line is your National Park Service Director." To which Udall, a grandson of John D. Lee of Lee's Ferry, responded, "We are working on that."

The congressional mandate of August 25, 1916, ordered the secretary of the interior to manage our national parks "by such means as will leave them unimpaired for the enjoyment of future generations." By no stretch of the imagination could these dams leave the Colorado unimpaired, and as far as Brower was concerned, Grand Canyon would be dammed over his dead body. Leading the Sierra Club and, by example, other conservation organizations, Brower organized a no-holds-barred campaign against the bureau and Congress.

In *Cadillac Desert*, Marc Reisner describes Brower's first weapon against the bureau: Jeffrey Ingram, a young mathematician with a passion for Grand Canyon. Ingram dissected the bureau's financial justification for the dams, and what he discovered was astonishing. The bureau planned an illegal attachment of all revenues from Hoover, Parker, and Davis dams in the late 1980s—money that should have reverted to the United States treasury. Worse, the new Grand Canyon dams would be so expensive (Bridge Canyon Dam alone would cost over a billion dollars) that, even with the illegal revenues, decades would pass before the bureau could afford any of its bizarre projects to augment the Colorado. During those decades the dams would create a net loss of water. Not building the dams, Ingram concluded, would result in more water and more revenues than building them. Ingram's analysis embarrassed the Bureau of Wreck-the-Nation and killed the more bizarre components of Dominy's Pacific Southwest Water Plan. But, surprisingly, it did not faze the CAP and its need for cash-register dams.

Brower's second weapon was publicity. The Sierra Club used every medium: books, mailings, films, and the most notorious newspaper ads ever published for conservation. The *Washington Post*, the *New York Times*, the *Los Angeles Times*, and the *San Francisco Chronicle* ran a series of full-page ads paid for by the Sierra Club warning Americans of the planned destruction. They countered the bureau's bizarre claims that dams would improve the Grand Canyon by asking, "Should we also flood the Sistine Chapel so tourists can get nearer the ceiling?" Other ads read, "Now only you can prevent Grand Canyon from being flooded for profit." and "Who can save Grand Canyon—An open letter to Stewart Udall."

Brower's tactics succeeded wildly. Fradkin reports that many eastern members of Congress received a sudden spate of letters and telegrams of protest greater than on any other issue during the Eighty-ninth Congress (and this was during the hottest phase of the Vietnam War). Reisner reports that 95 percent of the letters the bureau itself received were against the dams. Overnight, voting in favor of these two dams in Grand Canyon became a political liability.

Congress said no to the dams, but the cost of victory was high. The Internal Revenue Service counterattacked the Sierra Club by construing newspaper ads as lobbying and by redefining it as a lobbying organization to void its status as a valid recipient of tax deductible contributions.

The total cost was higher yet. The bureau still had to replace the lost cash-register dams that were vital to its plan to defy gravity to allow the CAP to rescue the absurd growth in central Arizona. Stewart Udall floated the Colorado River through Grand Canyon and became so opposed to the dams that, while Floyd Dominy was outside the country, he ordered them scratched from the bureau's plan. But, a pragmatist and a native Arizonan, Udall had to find a replacement.

In 1967 that replacement was revealed: the bureau would buy in to a huge coal-fired plant constructed in Navajoland, the Navajo Generating Station. In 1968, Lyndon B. Johnson signed the now-trimmer Colorado River Basin Project Act to pay for the Navajo plant and the Central Arizona Project. Reisner notes that, even after being trimmed, this act was still "the most expensive single authorization in history."

In the 1970s the Navajo Generating Station started belching smoke over the Four Corners region and Painted Desert, obscuring the view across Grand Canyon for three million visitors a year. Coal to feed it was strip-mined off sacred Hopi land on Black Mesa. More full-page ads by conservationists in 1971 likened this strip-mining to "ripping apart St. Peter's in order to sell the marble."

Yes, the cost of stopping the dams had been high, and doubtless the Central Arizona Project and the Central Utah Project will demand even greater sacrifices in the future. But before destroying even more of our besieged western ecosystems on the altar of real estate and beef sales, I think we should sacrifice the sacred cows.

The final cost of Brower's victory was Brower himself. In 1969 ten of the Sierra Club's board of fifteen directors voted to remove him as executive director. The board complained that Brower rode the Sierra Club like the Lone Ranger rode Silver, as a loner who believed he alone knew what was best for conservation in America—and who, by firing too many silver bullets, was fiscally irresponsible to boot. The fact that Brower had beaten the bad guys better than anyone else had done did not seem to count.

Brower was stunned, but he reentered the conservation arena, founding Friends of the Earth, a more radical and aggressive organization. Years later, the directors of Friends of the Earth usurped Brower's control, so he created still another organization, the Earth Island Institute, where, in his late seventies, he is even more serious about environmental defense than ever, applauding the "eco-defense" tactics of Earth First! with, "The environmental movement has gotten very drowsy, and I think Earth First! is giving it CPR."

And Floyd Dominy? Ironically, his twenty-five years with the Bureau of Reclamation ended right after the Sierra Club jettisoned David Brower. Marc Reisner notes that Dominy's weakness—numerous sexual indiscretions that he not only failed to conceal but actually flaunted—finally destroyed his career. When Richard Nixon, newly inaugurated as president, saw the FBI's thick file on Dominy, he ordered him fired. Dominy retreated to his wife and ranch in Virginia to raise cows.

The stars had faded. The Redwall cliffs were now etched clearly for 500 feet straight up. The smell of coffee mingled with the aroma of bacon sizzling in the Dutch oven and overpowered the odor of the algae now exposed by the bureau's vacillating Colorado. I rarely drank coffee and had long since given up bacon, but their aromas made Nautiloid Canyon smell like home.

Now the boats were tipping very close to the angle of discomfort. I glanced at Fabry. He was tucked into a down-filled bag that looked like it had been strapped to the last axle of a semi-trailer for a coast-to-coast run. He was still dead to the world. I lay back down to start my morning stretches to loosen the muscles in my lower back. In the middle of these T. A. yelled, "Eggs and hotcakes! Get 'em now or go hungry!"

I owed David Brower and his allies a debt. We all did. Were it not for him we would all be wearing scuba gear down here. To every boatman who rows the Colorado, he is a saint. His story is the ultimate song of victory for this river. But the saga of the Grand Canyon dams is not over. Even as I write, some Arizonans are pushing for the construction of the Bridge Canyon Dam near Mile 238. Why?

You can probably guess.

6

Major Challenges

e have boats that we can row downriver. With powerful engines we can blast upstream as well. Now, in the final decade of the twentieth century, we can also travel underwater, in the air, underground, and through the emptiness of space. Only one dimension still defies two-way travel: time. But if we could warp the fabric of time, I would pack my boat into the time machine and set the dial for the Colorado River in Grand Canyon in August 1869.

Why? To rendezvous with Major John Wesley Powell, the man who explored North America's most inaccessible terra incognita—stretches hundreds of miles long and a hundred wide in the canyons of the Green River and the Grand Canyon of the Colorado. Powell grabbed national headlines in 1869 when he revealed the secrets of the West's last great unknown region. His feat compared with Lewis and Clark's expedition across the continent, but for one detail: Lewis and Clark did not lose a man after embarking into the unknown. Powell did.

Even now, more than a century after he led his eight men down a thousand miles of the Colorado's unknown canyons, Powell has supporters who canonize him as the only man who could have pulled it off, and detractors as well, who excoriate him as a blowhard who got lucky in choosing his crew. I am haunted at times, wondering which opinion is more justified. Deciding is not easy. I need a time machine. And I am not alone in this. Virtually every guide I know would stampede into it to meet Powell and his heroic crew of 1869.

Digging beyond Powell's self-serving *Exploration of the Colorado River and Its Canyons* into the journals and accounts of the men who rowed those first boats down the Colorado opens a Pandora's Box. Many times while drifting downriver I have fantasized about what I might say or do were I to warp through time, row around a bend, and come upon that first Powell expedition. I might have only seconds before the cosmic singularity whipped me back to the future or the power source of my time machine ran too low. What would I ask? What would I tell? The right few words would have been lifesavers, ill-considered ones a tragedy. What warning would I give? What would *you* say?

During the Battle of Shiloh, a year after Powell enlisted in the Union Army, a Confederate minié ball shattered his right forearm. After the South surrendered, Powell, emaciated and missing his right arm, taught natural history and expanded his university field trips into Indian country, always going personally unarmed through the Rocky Mountains at a time when George Armstrong Custer was posturing as a military genius and General Phil Sheridan was voicing his policy: "The only good Indian I ever saw was dead."

In 1868 Powell decided to explore the Colorado River. His small government grant and private contributions were enough to pay a seventy-five-dollar wage to three hunters (for the entire expedition) but nothing to the six volunteers. Instead of a balance of scientists and outdoorsmen, Powell selected his brother Walter plus five trappers: Oramel and Seneca Howland, William Dunn, John C. Sumner, and William Hawkins (the last three were the "paid" hunters). He also recruited an Englishman named Frank Goodman and Americans Andy Hall and George Bradley, who said he "would be willing to explore the River Styx" if he could thereby escape serving the rest of his hitch in the army. Six of Powell's nine men had served in the Union Army.

On May 24, 1869, Powell's learn-as-you-go expedition set off from Green River, Wyoming, in three heavy oak boats—Whitehall haulers—21 feet long by 6 abeam by 26 inches deep. They were decked fore and aft for 5 feet with watertight compartments, between which sat two oarsmen who rowed and steered (no boat was steered from the stern with a sweep oar). The boats *Kitty Clyde's Sister*, *Maid of the Canyon*, and *No Name* were marginal in white water. No boatman would row them today. With ten months' provisions and crew, each boat carried 1,000 pounds of food, a hellish load for two men rowing challenging water—even if they knew how to row white water, which they did not—and a nearly impossible load once the boats had been swamped in rapids. Powell's fourth craft, his pilot boat, was 16 feet long by 4 abeam and built of pine and lightly loaded. He named it the *Emma Dean* (his wife's maiden name).

After a week on the Green, the rapids convinced Powell to order his men to portage their equipment and supplies, then line the boats down along the rapid's edge with ropes. Such lining would become routine but would remain brutal, painful work that would be repeated at 62 rapids. On day sixteen in the Canyon of Lodore, Oramel G. Howland captained the *No Name* on its last ride, a ride that would spawn the greatest controversy in the history of the Colorado River.

The *No Name* was swamped from a rapid they had just run, and too late its crew saw that Powell's lighter *Emma Dean* and the other two boats had pulled to shore. Powell's standing orders were to row the boats 200 yards apart, but the other captains agreed that neither Powell nor his boatman, William Dunn, had signaled to stop. The Howland brothers and Goodman struggled toward shore, missed, and entered "Disaster Falls" blindly. Their boat broached on a boulder, swamped, spun off, then was swept downstream to collide broadside with another rock, where the river snapped it in two. The Howlands and Goodman clung to the larger half until it cracked into a third boulder and was demolished. Then, without life jackets (Powell had procured only one life jacket—an inflatable rubber one for himself), they clawed their way to an island. "Thirty feet further down," Jack Sumner wrote, "nothing could have saved them, as the river was turned into a perfect hell of waters that nothing could enter and live."

Although Sumner rowed out in the *Emma Dean* and rescued the men above that rapid, the *No Name* and its contents were a total loss. In the first leg of their journey, with the tough stretches still ahead, the expedition had lost more than a quarter of its material assets and supplies in one fell swoop.

Powell admits being elated over the men's rescue and the only recoveries: all three barometers and two thermometers, plus a three-gallon keg of whiskey he had known nothing about. But he writes no more. Sumner writes that Powell angrily demanded that O. G. Howland explain why he had not landed. Lack of a signal, Howland responded. This incident would fester like a tropical ulcer.

The ten men then struggled for a week down the twenty miles of Lodore Canyon to enter more than 300 idyllic miles of easy—or dead flat—water. They hiked up the Uinta River to the Ute agency to mail letters and buy 300 pounds of flour. Frank Goodman, having lost everything in the wreck of the *No-Name*, quit Powell's crew. This alleviated Powell's shortage of boats and rations.

Despite the relatively easy water of Desolation Canyon, Powell's *Emma Dean* flipped, tossing Powell, Dunn, and Sumner into dangerous water. They lost two rifles, all three bedrolls, and a barometer. Finally the expedition found respite on the 120 miles of placid water in Labyrinth and Stillwater canyons.

One month and 300 miles after escaping the last rapid in Lodore, Powell's expedition arrived at the confluence of the Green and the Colorado at the head of Cataract Canyon to launch a "vigorous campaign." After a few days spent recaulking the boats and scouting the river from the canyon rim, on July 21 they set off again. Powell's *Emma Dean* flipped again in the first big rapid. Then, for a week, the expeditionaries struggled down Cataract Canyon, portaging and lining several rapids they considered the worst so far—though most were small compared to what awaited them in Grand Canyon. Twice again they stopped to saw new oars from driftwood to replace those lost from the *Emma Dean*. The men shot two desert bighorn—a welcome break from their monotonous fare of rancid bacon, musty flour, re-dried apples, a few beans, and black coffee. By now they had lost or consumed eight months of their ten months of provisions.

Below Cataract lay the quiet water of Narrow and Glen canyons, nearly 180 miles of liquid euphoria where they zipped downriver ten times faster. They floated past the Paria and future Lee's Ferry on August 5—the height of the monsoons. The same day they portaged Badger Rapid and then Soap Creek. Their camp at Soap Creek marked seventy-four days navigating 700 miles of river, most of them a first descent. The nine men were now hardened veterans whose river-running experience would not be surpassed until the twentieth century. At this point, however, they were also tired, malnourished, and nearly without boots and clothing—and even without a blanket apiece to cover themselves during the nightly storms. All but one of the wagon tarps they had used to construct shelters against the rain had long since gone down the river—and they had twenty-six days to go. Three days into Marble Canyon, Bradley wrote: "Have been in camp all day repairing boats, for constant banging against rocks has begun to tell sadly on them and they are growing old faster if possible than we are."

For six days Powell's men rowed deeper into Marble Canyon, portaging and lining House Rock Rapid and many of the Roaring Twenties. They arrived at the mouth of the Little Colorado (Mile 61.5 from Lee's Ferry) by August 10. Here Powell halted for three days to survey the region, which inspired him (years later) to pen his most famous lines on the unknown terrors ahead.

[August 13] We are now ready to start our way down the Great Unknown. Our boats, tied to a common stake, chafe each other as they are tossed by the fretful river. They ride high and buoyant, for their loads are lighter than we could desire. We have but a month's provisions remaining. . . .

We are three-quarters of a mile into the depths of the earth, and the great river shrinks into insignificance as it dashes its angry waves against the walls and cliffs that rise to the world above; the waves are but puny ripples, and we but pygmies, running up and down the sands or lost among the boulders.

We have an unknown distance yet to run, an unknown river to explore. What falls there are, we know not; what rocks beset the channel, we know not; what walls rise over the river, we know not. Ah, well! we may conjecture many things. The men talk as cheerfully as ever; jests are bandied about freely this morning; but to me the cheer is somber and the jests are ghastly.

Although poetic enough to raise the hairs on the back of the neck of any boatman who has challenged an unknown river, Powell's account clashes with what George Bradley wrote in his diary (oddly, Bradley kept his journal so secretly that Powell did not know it existed). Bradley said they had camped for three days along "a loathsome little stream, so filthy and muddy that it fairly stinks." Sumner agreed, calling it "as disgusting a stream as there is on the continent. . . . Half of its volume and 2/3 of its weight is mud and silt." More revealing, instead of a "cheerful" crew, Bradley wrote that "The men are uneasy and discontented and eager to move on. If the Major does not do something soon I fear the consequences, but he is contented and seems to think biscuit made of sour and musty flour and a few dried apples is ample to sustain a laboring man."

Oblivious or not of his men's true feelings, Powell had to continue into the "Great Unknown." Fifteen miles downstream they entered First Granite Gorge, jagged walls of metamorphic schists and gneisses nearly two billion years old and a mile deep, locking them onto forty of the narrowest and steepest miles in the Canyon and the worst rapids on the river (except for Separation and Lava Cliff rapid 150 miles downstream). A few were impossible to portage. Bradley wrote: "no rocks ever made can make much worse rapids than we now have."

But the river did get worse. "This part of the canyon is probably the worst hole in America, if not in the world," Sumner later wrote. "The gloomy black rock of the Archean Formation drive all the spirit out of a man. And the excessive drenching and hard work drive all the strength out of him and leave him in a bad fix indeed. We had to move on or starve."

The expedition became a single-minded struggle to survive. They portaged when they could but ran scared, almost willy-nilly, where they could not. Powell wrote later of August 15, "The boats are entirely unmanageable."

The next day they stopped at Bright Angel Creek to saw replacement oars again and to dry their rations—ten days' supply. Three days later the *Emma Dean* flipped a third time. After a nine-day nightmare of gigantic waves and grueling portages descending First Granite Gorge, they raced downriver, rowing thirty-five miles in a day. Now on less than half rations, hunger was a constant companion. But the worst was yet to come.

Ten days after being reduced to ten days' provisions, they spied an Indian garden, the first in hundreds of miles, probably at Indian Canyon (Mile 206.5). They rushed into it, grabbed a dozen squashes, then hightailed it back to their boats to escape like desperados.

The men made excellent time from Indian Canyon, stopping before noon the next day to scout the bad rapid above Mile 240 (now submerged under Lake Mead). George Bradley wrote secretly in his journal: "we came to the worst rapid yet seen. The water dashes against the left bank and then is thrown against the right. The billows are huge and I fear our boats could not ride them even if we could keep them off the rocks. The spectacle is appalling to us. . . . This is decidedly the darkest day of the trip but I don't despair yet. I shall be one to try to run it rather than take to the mountains."

Here occurred Powell's most notorious incident, the one that most haunts professional boatmen. Hours of scouting from both banks revealed that the rapid was not only among the worst so far but also that it was impossible to line or portage. Finally Powell announced that they would run it in the morning. Oramel Howland told Powell that it was madness to proceed and that his brother Seneca, William Dunn, and he had decided to hike out north into Utah no matter what Powell did.

Powell reckoned eighty or ninety miles of river remained between them and the Rio Virgin—the second half known to be easy water. He further estimated it was seventy-five miles overland to the nearest Mormon settlement. He showed Howland his figures, but Howland remained obdurate.

"At one time I almost conclude to leave the river," Powell described his quandary of August 27. "But for years I have been contemplating this trip. To leave the exploration unfinished, to say there is part of the canyon which I cannot explore, having nearly accomplished it, is more than I am willing to acknowledge, and I determine to go on."

This was where I would aim my time machine—and my entire trip. August 27, 1869. Evening. Mile 240. Separation Rapid.

Imagine rowing to shore here. Yellow fiberglass oars whipping across the smooth, silt-laden water above the rapid to propel our bright yellow inflatable boats rigged with heavy aluminum frames, decking, and boxes. In these boxes, in the drop bags beneath our decks, and in our 172-quart coolers there is still more food than we can eat. In fact, the greatest single curse here is gaining weight from overeating the rich food—despite the many hikes and all the rowing. Too much beer. Ice cold beer.

On shore are tied three wooden boats, faded, gouged, scratched, cracked, splintered, patched, leaky, but strangely buoyant—they are virtually empty of cargo. The oars are crude, hewn from driftwood poles. The decks are innocent of equipment. The bowlines of hemp look as frayed as if they had been dragged over a hundred miles of sandstone.

Most stark and telling is the crew. Short, gaunt, sun-baked, silt-hewed, unshaven, wearing nearly nothing in the hundred-degree heat but a ragged tatter of shirt and a belt securing a Bowie knife. And here and there a revolver. Not one pair of britches is visible. The bearded, one-armed man down by the rapid is naked but for a rubber horse-collar life jacket. The universal scowls and calloused fists rubbing stubbled chins reveal that this rapid has them worried. Down next to the roar of the rapid, each man wishing to make a comment on how to meet this challenge is forced to yell. But they are mostly silent.

We row to the sandy beach upstream from the rapid and its roar and leap out to tie nylon bowlines to boulders.

"Who the hell are you?" Billy Hawkins asks, looking up from a cluster of unbaked biscuits of unleavened dough lined up near his campfire. Nearby is a flour sack, now nearly empty. It appears that he is about to bake their entire supply of flour. Instead of a camp, all this crew has is this camp fire, biscuits, a battered pot of coffee, and a couple of bedrolls. Startled, but still a man who knows that hospitality is the ultimate propriety in the West, Hawkins offers a rough hand in greeting. I take it. Crumbs of dough grind between us to drop onto the fine sand.

"We're river runners . . . from the future," I say, knowing it sounds absurd. "I have an important message for you. Life and death. But only a few minutes to tell you what you need to hear. Take me to Powell."

Hawkins stares into my face, then looks hard at my boats and at the people dressed in cotton T-shirts, nylon shorts, and sunglasses milling around them. Cameras are popping out everywhere. Then he looks back at me. "You're what?"

"Commercial river runners," I tell him. "My name's Michael Ghiglieri. I'm leading this trip. It's . . . it's . . ." It's hard to explain this in a way he would believe. "People now go on these trips for fun, for sport. They hire us to guide them safely. We're from the late twentieth century. I know this is hard to believe. By the way, I've got some food I can give you. Even cold beer."

Hawkins takes me to Powell, who is astounded. He shakes my hand and now looks even more worried than when he was staring at the rapid.

"I've got a problem," Powell finally admits later, after we hike away from the roar and I have persuaded him that we are real. "Three of my boys are deserting the expedition. They want to hike out."

"Why?" I ask. I thought I had this mystery solved, but if I was wrong, I wanted to know. I want to know very badly. I might be able to help, but it depends on what was really going on here.

The evidence shouts that what really happened at Separation Rapid started eighty-one days earlier at Disaster Falls. Robert Brewster Stanton, who in 1890 commanded the second successful navigation of Grand Canyon, stumbled across the first clue by accident—a chance meeting with Jack Sumner in Glen Canyon as Stanton was beginning his own survey. Sumner, boatman on the *Emma Dean*, has generally been credited with being Powell's greatest asset. Sumner even told Stanton that it was he who convinced Powell in 1867 that the Colorado should be explored. At any rate, until this meeting with Sumner, Stanton had idolized Powell, having read his 1875 book, he said, "with an almost worshipful reverence . . . the most fascinating story I had ever read . . . of one of the bravest exploits ever known anywhere."

One can imagine Stanton's reaction when Sumner offered his own succinct assessment of Powell's book: "There's lots in that book besides the truth."

This triggered an apostasy in Stanton, who sought out Sumner again, as well as every other survivor of Powell's trips, to dig for the truth. Stanton's obsession drove him to hire a stenographer and notary public to record and witness his interviews (he ultimately compiled a tome on the Colorado so thick that no publisher would touch it). For Stanton the greatest issue in Powell's exploration was what had transpired at Separation Canyon. For me, and most boatmen, the same holds true.

In 1907 William Hawkins, a boatman on *Maid of the Canyon* and camp cook, gave Stanton his account (which was reported in Stanton's *Colorado River Controversies*). After saying that "no man living was ever thought more of by his men up to the time he wanted to drive Bill Dunn from the party," Hawkins explained:

> The trouble with the Howland boys began away back at
> Disaster Falls, where their boat was lost, but with Dunn it
> began [in Cataract Canyon] only a few weeks before he left
> the party. At noon one day when the boats were being let over
> a bad place, Dunn was down by the water's edge with a
> barometer, taking the altitude. He was also assigned the post
> to look after the rope fastened to the boat and held by Sumner
> and the others. By some means Dunn was thrown into the
> river, but he caught the rope and finally got out. In this he got
> wet a watch that belonged to the Major. At dinner the same
> day Major Powell told Dunn that he would have to pay for the
> watch or leave the party, which was impossible at that point.
> Dunn told him a bird could not get out of that place, thinking
> the Major was joking, but all of us were very quickly convinced
> that every word the Major said was meant. Dunn said he
> could not leave then, but that he would go as soon as he could
> get out. The Major then said he would have to pay one dollar
> a day for his board until such time as he could get out of the
> canyon. The rest of us sat listening as we ate our dinner. As
> Sumner was the oldest of our crowd [he was twenty-nine],
> that is, the two Howlands, Dunn, Hall, and myself, we
> naturally looked to him as our spokesman.

In 1919 Hawkins expanded this account in a talk with William Wallace Bass, a pioneer prospector and tour guide in the Canyon at the turn of the century, by explaining what happened next:

The next day we had some very bad rapids, so bad that it was necessary to let the boats around some large rocks. In order to do this, and as Dunn was a fine swimmer, the Major asked him to swim out to a rock so the boat would swing in below. He made the rock all O.K. and was ready to catch the rope which was supposed to be thrown to him, so he could swing in the boat below, but the Major saw his chance to drown Dunn, as he thought, and he held the rope. That was the first time he had interfered in the letting of the boats around bad places, and the rope caught Dunn around the legs and pulled him into the current and came near losing the boat.

But Dunn held onto the rope and finally stopped in water up to his hips. We were all in the water but the Major and Captain [Powell's brother]. Dunn told the Major that if he had not been a good swimmer he and the boat would have been lost. The Major said as to Dunn that there would have been but little loss. One word brought on another, and the Major called Dunn a bad name and Dunn said that if the Major was not a cripple he would not be called such bad names.

Then Captain Powell said he was not crippled, and started for Dunn with an oath, and the remark he would finish Dunn. He had to pass right by me and I knew that he would soon drown Dunn, as he, so much larger could easily do. He was swearing and his eyes looked like fire. Just as he passed I caught him by the hair of his head and pulled him over back into the water. Howland saw us scuffling and he was afraid Cap would get hold of my legs. But Dunn got to me first and said, "For God's sake, Bill, you will drown him!" By that time Howland was there and Cap had been in the water long enough and Dunn and Howland dragged him out on the sand bar in the rocks. After I got my hold in Cap's hair I was afraid to let go, for he was a very strong man. He was up in a short time, and mad! I guess he was mad! He cursed me to everything, even to being a "Missouri puke." I wasn't afraid of him when I got on dry ground. I could out knock him after he was picked up twice.

He made for his gun and swore he would kill me and

Dunn. But this talk did not excite me. As he was taking his gun from the deck of the boat, Andy Hall gave him a punch behind the ear and told him to put it back or off would go his head. Cap looked around and saw who had the gun, and he sure dropped his. (From Stanton's *Colorado River Controversies*)

Jack Sumner, whom Stanton admired as "a true frontiersman of the highest type . . . quiet and generous, and yet with a temper and spirit that knew no bounds when he was treated unjustly by others," also gave his account to Stanton:

after having had a spat with Howland in the forenoon, Major Powell at the noonday camp informed Dunn that he could leave the camp immediately or pay him fifty dollars a month for rations. . . .

As that little statement raised me to a white heat, I interposed and said that if any one voluntarily wished to leave, he could do so, but that no one could be driven from the outfit. Major Powell informed me that he was talking then, and commanding the expedition. I told him that he could talk all he pleased, but that he must cease then and there, his abuse of Howland and Dunn. Walter Powell tried a bluff and was immediately called to settle, as there was a pretty little sand bar just about long enough for Colt's forty-fours. He replied that he had no arms, and I told him to take his choice of my pistols and choose his distance. He did not accept the proposition. After that little episode, everything went as smooth as with two lovers after their first quarrel and make-up. Major Powell did not run the outfit in the same overbearing manner after that. At a portage or bad let-down he took his geological hammer and kept out of the way.

Hawkins and Sumner agree that the Howlands and William Dunn left the expedition not from cowardice but due to the friction between Powell and them, especially between Oramel Howland (the trip cartographer) and Powell and also Walter Powell, whom all the men considered demented. Even Powell revealed, between the lines, that personality conflicts split his expedition.

On the morning of August 28, both Howlands and Dunn took their share of Hawkins's biscuits, two rifles, a shotgun, ammunition, Powell's duplicate records (in haste divided unequally so that each bundle contained duplications and omissions), a $650 barometer, a letter to Emma, and Jack Sumner's silver watch, which was to be given to his sister. Before leaving, the three hikers helped carry the two oak boats to the best entry point into the long rapid. Powell abandoned the *Emma Dean*. From a crag the Howlands and Dunn watched Hawkins and Hall row *Maid of the Canyon* through Separation Rapid unscathed. Sumner, Bradley, and Walter Powell then rowed *Kitty Clyde's Sister*. Powell later wrote in his book: "Although it looked bad from above, we had passed many places that were worse." He added:

> We land at the first practicable point below, and fire our guns,
> as a signal to the men above that we have come over in safety.
> Here we remain a couple of hours, hoping that they will take
> the smaller boat and follow us. We are behind a curve in
> the canyon and cannot see up to where we left them, and so
> we wait until their coming seems hopeless, and then push on.

The six ran several easy rapids and then, at Mile 246, were gripped again at Lava Cliff Rapid, allegedly the worst rapid on the Colorado between Wyoming and the Sea of Cortez. From the top of the cliff they tried lining the first boat with Bradley aboard but ran short of rope. Powell ordered another length to be brought, but meanwhile the boat whipped back and forth, surfing in the current and smacking against the cliff. Afraid it would be smashed to kindling, Bradley drew his Bowie knife and stumbled forward to sever the bowline. As he reconsidered, the stempost broke free, sending the boat zipping into the rapid, out of control. Powell, Hawkins, and Hall jumped into the second boat to rescue Bradley, but their boat swamped immediately. Bradley had grabbed the oars to pilot himself through the pounding waves into an eddy. Then he turned the tables on his would-be rescuers by helping them bail their swamped boat.

By noon the next day (August 29, 1869), only thirty hours after leaving the Howlands and Dunn on their crag, the six rowed past Grand Wash Cliffs, the end of Grand Canyon. They had just made history.

The next day they reached the mouth of the Rio Virgin, where the Powell brothers ended their ninety-nine-day, thousand-mile journey. They still had ten pounds of flour, fifteen pounds of dried apples, and nearly eighty pounds of coffee. The men continued downriver.

But what of Oramel and Seneca Howland and William Dunn? It should be no surprise that Powell's detractors consider this episode the greatest indictment of the man. Powell did dedicate his book to all nine of his crew by name as "noble and generous companions." After all, at least three times they had saved Powell from his own lethal mistakes. But Powell devotes little of his narrative to their fate. Twenty-seven pages pass before we read a solitary sentence: "The three men who left us in the canyon last year . . . met with Indians . . . and were finally killed." Then we wait another dozen pages until Powell is in Shivwits country, meets the alleged killers, and relates their story (as translated to him by Mormon scout Jacob Hamblin):

> They came upon the Indian village almost starved and
> exhausted with fatigue. They were supplied with food and put
> on their way to the settlements. Shortly after they had left,
> an Indian from the east side of the Colorado arrived at their
> village and told them about a number of miners having killed a
> squaw in a drunken brawl, and no doubt these were the men;
> no person ever came down the canyon; that was impossible;
> they were trying to hide their guilt. In this way he worked
> them into a great rage. They followed, surrounded the men in
> ambush, and filled them full of arrows.

I find it highly revealing that in 400 pages Major Powell expended only a dozen sentences on the fate of the Howlands and Dunn and never expressed regret at their having been murdered. Maybe he felt that deserters deserved whatever fate was dealt

100

them. Unfortunately, it is impossible to read between the lines of Powell's original diary entries for that day (August 28): "Boys left us. Ran rapid. Bradley boat. Make camp on left bank. Camp 44."

How honest was Powell? Sumner's account of the events immediately following their running Separation Rapid clashes with Powell's account:

> We waited about two hours, fired guns, and motioned for the men [the Howlands and Dunn] to come on, as they could have done by climbing along the cliffs. The last thing we saw of them they were standing on the reef, *motioning us to go on,* which we finally did. If I remember rightly, Major Powell states it was not as bad as it looked, and that we had run worse. I flatly dispute that statement. At the stage of water we struck, I don't think there would be one chance in a thousand of making it by running the whole rapid. (Italics mine.)

Powell states that after the six ran Separation the Howlands and Dunn were hidden from view by a bend in the river. In fact, no bend exists for two miles. Sumner, Bradley, and Hawkins agree that all nine men were in view of one another until the three waved them on, by doing so clearly stating that they chose not to travel with the men, not that they feared the rapid, which, in any case, they simply could have walked around at this point.

Weighed against the accounts of the men who remained faithful to him to the end, Powell's self-serving fictionalization in his journal suggested to Stanton and others that the Howlands and Dunn left because of Powell himself.

What do I mean by *fictionalization?* In *The Exploration of the Colorado River and Its Canyons* (published under congressional pressure in 1875), Powell wrote an exciting and colorful personal account of his explorations of the Colorado but one for which he has been criticized for errors, hyperbole, and for some fictional episodes added for the sake of drama. For example, Powell noted that Redwall Cavern in Marble Canyon could comfortably sit 50,000 people. The next day he effused over the glories of a "wall set with a million brilliant gems," which he named Vasey's Paradise. The errors? Vasey's Paradise is a mile upstream from Redwall Cavern (Powell ran this section twice and should have known), and even 5,000 people in Redwall Cavern would be jammed like sardines. Powell also exaggerated the worst dangers. He described Sockdolager Rapid, for example—one of the few rapids they were forced to run because a portage was impossible—as descending "perhaps 75 or 80 feet in a third of a mile." Actually it drops 19 feet, and Powell's own journal entry on Sockdolager—"Chute ½ mile long. Fall 30 ft. probably, huge waves"—should have prevented him from exaggerating the descent by some 50 feet.

Powell also beefed up his account of his 1869 exploration by including events from his second trip in 1871 (aborted at Mile 143.5) as if they had occurred on the first, thus creating a misleading and partly fictional account that further obfuscates the truth. He used journals written by his second crew but gave no credit to their writers or to the crew who actually performed in these borrowed incidents. Powell erased their existence for the sake of a smoother narrative. He also borrowed heavily from his brother-in-law A. H. Thompson's diary of the second trip and, according to Stanton (who discussed this issue with both Powell and Thompson), discouraged Thompson from revealing any of the observations from his personal journal or scientific findings to the public. Until Powell's dying day, Stanton notes, he generally referred to his explorations of the Colorado River as having occurred in 1869, period. The men from his 1871 trip felt betrayed.

Powell's semifictional book did work wonders for him. Combined with his vision and tenacity, it hurled him into the political arena. His treasure trove of ethnographic data on Indians earned him the directorship of a department he created in the Smithsonian Institution, the Bureau of Ethnology, and led to the two-volume *Handbook of American Indians,* the best ethnographic source on the Indians until the Smithsonian's effort to produce an updated handbook in twenty volumes. In 1881 Powell became the director of the U.S. Geological Survey, where he espoused, unpopularly, land use based on a sustainable yield within the constraints of local water supplies, confining development to within a river basin and prohibiting diversions from one basin to another to sell land in the second.

The case on the Howlands and Dunn is closed. Or is it? Powell states that they were killed by the Shivwits. Powell himself did not speak Shivwits. The Indians' frank admission (a surprising one, to put it mildly) that they had killed all three men came to Powell through the interpreting services of Jacob Hamblin. Concerning this, Sumner is skeptical:

> While we repaired the boats [a few hours after running Lava Cliff Rapid at Mile 246] the boys discussed the conduct and the fate of the three men left above. They all seemed to think the red bellies would surely get them. But I could not believe that the reds would get them, as I had trained Dunn for two years in how to avoid a surprise, and I did not think the red devils would make open attack on three armed men. But I did have some misgivings that they would not escape the double-eyed white devils that infested that part of the country. Grapevine reports convinced me later that that was their fate. . . .
>
> I heard about two months afterwards, while at Fort Yuma, California, that they [Mormon searchers] brought in the report that the Howland brothers and Dunn came to an Indian camp, shot an Indian, and ravished and shot three squaws, and that the Indians then collected a force and killed all three of the men. But I am positive I saw some years afterward the silver watch that I had given Howland. I was with some men in a carousal. One of them had a watch and boasted how he came by it. I tried to get hold of it so as to identify it by a certain screw that I had made and put in myself, but it was spirited away, and I was never afterwards

able to get sight of it. Such evidence is not conclusive, but all of it was enough to convince me that the Indians were not at the head of the murder, if they had anything to do with it.

Were the Howlands and Dunn murdered by Indians or by whites—and if by whites, why? One might guess that if we do not know by now, we never will. But this may be wrong. Throughout the West, murders were committed by whites who blamed their crimes on Indians. Some whites even dressed as Indians while committing murder. The most sophisticated went so far as to create secret military societies whose modus operandi included conducting organized raids and killings in the guise of Indians—as was still the case in southern Utah in 1869, a year of continued prejudice against Mormons and of threats of federal investigation concerning the as-yet-unsolved Mountain Meadows Massacre. Although it may come as a surprise, investigation into the disappearance of Oramel and Seneca Howland and William Dunn is still active, and Indians are not the suspects.

So much for the myth of Major Powell and his merry men.

Again I caught myself fantasizing about that time warp—about meeting Powell and his crew at Separation. What would I say to them? To Powell, "Try harder to keep Howland and Dunn with you. Apologize to them. Be a leader. Explain how important they are to the expedition. Tell them what you owe them, that you appreciate them, that you *need* them." And to the Howlands and Dunn? "Don't quit this expedition. In thirty hours by boat you'll escape Grand Canyon and win a place in history. If you hike out instead, you will be treacherously murdered. Go downriver even if you have to crawl on your hands and knees." Would they listen to a stranger? Would I have time to explain everything?

It's only a fantasy. But I still wonder, what *would* I tell them?

Finally, Grand Canyon. We had reached the confluence with the Little Colorado, the "foot" of John Wesley Powell's Marble Canyon. When Powell explored the river, he gave the canyon that begins at Lee's Ferry the name of Marble Canyon—despite its total lack of marble. But Marble Canyon had to end somewhere to make room on the map for the already named Grand Canyon. The only place it could "end" is here at the confluence with the Little Colorado River at Mile 61.5. Hence, even though no break or geological change exists in the Canyon between upstream and downstream of the Little Colorado—or anywhere else between Lee's Ferry and the distant Grand Wash Cliffs—Powell's name stuck.

The Painted Desert was now at the peak of its dry season; no surface water flowed along the upper two hundred miles of the Little Colorado. Its sinuous canyon flanked a dry bed of fissured and curled clay. It was a dead ringer for a desiccated wadi bordering the Empty Quarter of the Great Arabian Desert. It looked as dead as Mars. But here at its confluence with the Colorado, the Little Colorado was abruptly resurrected in a flowing, opalescent turquoise that seemed an illusion, like water that had originated on some other planet.

Despite its unearthly appearance, this flow was instead very earthly water. A dozen miles upstream from us it bubbled from artesian springs in the Redwall Limestone and from other seeps and springs in other formations down to the Tapeats Sandstone, a rock that replaces the rainbow-hued beds of Bright Angel Shale at river level a couple of miles upstream from the confluence here. Its turquoise hue is due to its rich concentration of calcium carbonate and other salts, and to the microscopic botanical garden of specially adapted algae that thrive in it. The color is so surprising that a first glimpse of the Little Colorado makes people gasp.

Even so, the Tapeats Sandstone itself affected me more. For me, reaching the Tapeats is like meeting an old friend. It's hard to explain. Size has nothing to do with it; its mere two- or three-hundred-foot thickness is not spectacular. But it remains the one formation in Grand Canyon that fires my atavistic neurons.

Why? Because Tapeats is a coarse sandstone sculpted by erosion and carved between its unconformities into ledges and benches. It is the only rock other than Redwall among the more than two dozen formations in the Canyon whose pattern of erosion results in natural cliff shelters. These offer one of the rarest commodities down here: shade. They also provide shelter from monsoonal storms, some of which are merciless.

Because of the sanctuary it provides, for at least 4,000 years hundreds of overhangs in the Tapeats throughout Grand Canyon have been used as hunting camps and shelters—first by Paleo-Indian hunters of the Desert Archaic culture, next by the Basketmaker cultures, then the Anasazi, and then by the Paiute hunters who walked atop the footprints of the vanished Anasazi (and now by me as I squatted in the shade and gazed across the turquoise flow rippled by the cavorting of life-jacketed dolphins). Many of these overhangs in the Tapeats cliffs remain obvious as camps; scattered atop and buried in their sandy floors are stone tools, projectile points, pottery sherds, mummified bones of bighorn sheep and deer, charcoal, partly buried caches of firewood, and sometimes even mats woven from yucca fiber. Stone Age condominiums, partly furnished and rent free.

When, as now, I sit in one of these cliff shelters, a Johnny-come-lately in the human history of this region, I feel the presence of these early hunters. It is almost as if they might stroll around a bend in the sandstone with the day's harvest of bighorn haunches dangling from one shoulder, a quiver of atlatl darts strapped across the other, a throwing stick in one hand, eagle feathers fluttering from tresses of raven hair, and grins of anticipation on sun-browned faces soon to be smeared with sheep fat. Sometimes while squatting here in the shade I feel so connected to these vanished primeval hunters that I start wondering where the best places might be to ambush a desert bighorn or mule deer and then find myself planning strategies for the hunt. These ledges are little time machines.

One nice ledge across the turquoise water from me now, on the west shore, was used as a camp by hunters perhaps three to four millennia ago. Most likely they were Paleo-Indians of the Pinto complex of the Desert Archaic culture who had hunted bighorn and deer here with their atlatls (the bow and arrow did not appear in this part of the Southwest until about the first century A.D.). These Paleo-Indians were enigmatic. They appeared here at least 5,000 years after the earliest Paleo-Indians reached North America from Mongolia, people of the Clovis and Folsom cultures who were such adept hunters that they helped push most of the giant mammals of the Pleistocene to extinction.

Those earliest big-game hunters trekked across the Bering Land Bridge, which existed only because the massive glaciers of those times locked up so much of the planet's water that sea levels worldwide dropped by at least 500 feet. They arrived in Alaska about 12,000 years ago (although evidence from Tom Dillehay's controversial dig in Monte Verde, Chile, suggests it may have been a thousand years earlier). Paleoecologist Paul Martin notes that in the first two thousand years or so after their arrival, fully thirty-three of the forty-five genera of large mammals in North

America vanished. This meant the abrupt extinction of 73 percent of all large mammal genera that were here. Sabertooths, giant lions, cheetahs, giant ground sloths, mammoths, mastodons, tapirs, horses, camels, gomphotheres, giant beavers, capybaras, spectacled bears, and many other giant creatures—each ignorant of the danger posed by the puny but lethal new two-legged primates trekking across their landscape—were swept off the continent as if by a tidal wave. In North America 10,000 years ago, Noah's Ark sank like the *Titanic*.

Among the few large survivors were mammals that themselves had migrated during the Pleistocene to North America from the Eurasian continent with an instinctive fear of humans: deer, moose, elk, bison, bighorn sheep, and grizzly bear. The fauna of South America fared even worse against these hunters; 80 percent of the large-mammal genera there became extinct. Atmospheric temperatures rose 12.5° F in a mere twenty years around 10,700 years ago to drop the curtain on the Pleistocene. Since then the temperature has risen another 12.5° F since the start of the Holocene epoch, the sudden spell of warming and drying that is still with us.

After this, the descendants of the Clovis and Folsom hunters, including people who preceded the Desert Archaic culture here, abruptly found life a greater challenge than it had been in the cornucopia of the Pleistocene. Now gone from the Canyon were the widespread Harrington's mountain goats, camels that lived along the river, a small horse (*Equus powell*), and the occasional giant sloth, mammoth, and shrub ox. These later hunters were forced to rely more on plant foods and to refine their hunting techniques to capture the surviving—and more wary—prey.

In many of the Redwall caves here in the Canyon 4,000 years ago, these hunters of the Desert Archaic culture buried, under stone cairns, figurines woven of split willow resembling deer and bighorn pierced with spearlike projectiles. Were these magic talismans for hunting? Certainly hunting was a challenge, especially here. Even so, millennia of hunters stalked the Canyon until the last Anasazi vanished.

I stared across the turquoise river at that cliff shelter. This has been a busy place. A recent excavation there unearthed artifacts of the Desert Archaic culture. On top of these are scattered stone tools and potsherds, some from the more recent Anasazi and the foundation of a stone dwelling built within the shelter of the Tapeats. In the early 1880s—only a dozen years after Powell and his men shoved off from here into the Great Unknown, and before any archaeologist got a chance to poke his trowel into this site—a prospector named Ben Beamer found his way here, liked the place, and eventually built a stone cabin atop the ruins left by the Anasazi. His cabin still stands. And today rusting iron tools, shards of glass tinted blue by the desert sun, and bits of shattered nineteenth-century crockery commingle with tools of flaked chert left by Ute hunters a few centuries back, Anasazi pottery and other artifacts a thousand years old, and broken stone tools of the Desert Archaic hunters almost contemporary with the First Dynasty in Egypt—an archaeological lasagna.

The Tapeats Sandstone intrigues not only me; it also draws the Hopi into the Canyon on sacred pilgrimages for reasons tied to its geological history. Tapeats is the oldest of the Paleozoic formations in the Canyon, a sandstone deposited about 570 million years ago in an energetic portion of the seas when this continent was 10,000 miles east-southeast of here. Its rough sands were the home of trilobites—one of the earliest organisms with legs in the history of life on earth. Their tracks, plus massive tangles of petrified worm burrows, crawl into eternity here in exposed layers of Tapeats.

Frequently, it seems, the Tapeats Sea either evaporated or regressed slowly to leave the sandstone beds heavily charged with sea salts. These deposits still seep out to form stalactites of nearly pure table salt. Two miles downstream from here they are so plentiful that they were harvested regularly by Hopis, who used the salt at home for both cooking and ceremonial purposes. The south side of the Canyon there, known as the Hopi Salt Mines, is considered sacred and is off-limits to river runners. A few Hopis still hike the Salt Trail from the village of Old Oraibi into the canyon of the Little Colorado seven miles upstream, then continue downstream to gather sacred salt for ancient ceremonies. Even many Hopis who have never been here can describe the confluence of the Little Colorado—they learned it by heart from hearing the traditional description during ceremonies in their sacred kivas.

I stared at the turquoise water. Even after all these years I am smitten by this stream's color during the dry season. The Little Colorado really is an opalescent turquoise—an impossible color in nature. No one believes it.

Besides being unbelievably beautiful, the Little Colorado is nearly twenty degrees warmer (68° F) than the Colorado. This warm stream, combined with air temperatures in the shade skidding past 100° F, evokes the hedonist in most of us. So, instead of being what Jack Sumner called "as disgusting a stream as there is on the continent," the Little Colorado now seems too good to be true. When Powell camped here during the August monsoons, the Little Colorado flowed thick with clay and other sediments. When it is swollen with runoff from pounding storms eroding the sedimentary formations of the Painted Desert to the east, one dunking in the Little Colorado will paint you brown. It is disgusting, unless mud turns you on. Soon this little river would again be as Powell, Sumner, Bradley, and the rest of the crew of 1869 saw it, but today, floating and cavorting in it and running its translucent little rapids wearing a life jacket (strapped around the butt instead of worn normally), makes everyone wonder whether leaving childhood behind had, after all, been an intelligent decision.

Three miles downstream from this gem-quality water, beyond the Hopi Salt Mines and the Land of a Thousand Eddies (transformed by afternoon winds to a rowing exercise from hell), is one of my favorite hikes in the upper Canyon.

I tied three half hitches in my bowline around a thick tamarisk. It may seem silly, but I resent these little salt cedars, so even though this tree helped me, I felt a mild disgust for it. Tamarisks do not belong here. They are exotics, the photosynthetic equivalent of German cockroaches. Our Department of Agriculture imported two or three species of tamarisks from the eastern Mediterranean around 1900 to line diversion canals and irrigation works of the Southwest to check bank erosion. Even though native riparian vegetation had been stabilizing stream banks here for thousands of years, Agriculture had looked elsewhere. Probably the irrigators were in too big a hurry to use slow-growing native plants.

The tamarisk story soon became a familiar one: the exotic panacea turned monster. When the introduced tamarisks matured, they sucked so much water out of the canals—every hour as much water as the weight of their canopy—that they more than canceled their value as preventers of erosion. Forty years after Agriculture planted the tamarisks, the Bureau of Reclamation tried to yank them all out. Then it spent millions to poison them, and failed. By then, feathery little tamarisk seeds—a quarter *billion* seeds in a single season from a big tammy—had blown upstream for miles. And almost nothing would eat tamarisk; even beaver turned their noses away. Today tamarisks have climbed up all the southwestern rivers to about 5,500 feet in elevation. Here they are the vegetable kings of the Canyon, hogging the Bureau of Reclamation's new high-water line. I hated them. Even so, I tied my boat to this one. I was in a hurry and it was convenient.

A few feet ahead, a collared lizard flowed up a block of Tapeats like a streak of light. He halted atop it to swivel his ferocious head and nail me with a suspicious gaze. Then he executed three jerky push-ups, a territorial warning normally aimed at other lizards. Despite his wrong assumption about what would intimidate me, he was impressive: designed for speed like some tiny predatory dinosaur now dust for sixty-five million years. But this one was alive and pumping itself up and down in a technicolor warning not to encroach on his territory. His black and white collar above a speckled tan back looked like war paint. Abruptly he turned and vanished beyond the lip of the rock. He raced across the graveled slope on his hind legs, uncannily like that vanished dinosaurian cousin. As quickly as he had appeared in this shimmeringly hot canyon, he vanished, miragelike.

Crunching gravel echoed off the close walls of swirled sandstone of Carbon Canyon (Mile 64). These walls climb hundreds of feet and are pocked everywhere with ledges and niches. In its bottom are scattered house-sized blocks of sandstone that have tumbled from the walls to lie helter-skelter, some leaning atop one another like the abandoned building blocks of a titan's child. A few of these giant, angular boulders provide rooms of shade under their bulk, oases of coolness in a landscape baking in solar radiation. The canyon ahead is so narrow and deep that it seems a special route to some unknown land, a place where normality has been suspended in some way.

Eleven hikers followed me into this labyrinth. Fabry took the rear to guide those who lagged behind and lost the route. We wound our way up through the upper Dox Sandstone, a 1.1-billion-year-old rock visible for the first time a mile upriver.

In single file we clambered up short cliffs, squeezed and oozed and trembled ascending chimneys between house-sized boulders tumbled from the Tapeats, and climbed several dry waterfalls typical of most tributary canyons here. Two-thirds of the way up Carbon we stepped above the gently sloping beds of Dox to pass the Great Unconformity—great because Tapeats Sandstone is only 570 million years old, while the Dox it sits on is more than a billion years old. Downstream fourteen miles the Tapeats sits instead upon Vishnu Schist nearly two billion years old. The unconformity between them is only an interface but one that represents a gap in time of roughly 1.4 billion years—nearly one third of the planet's history. This gap was created long before the Tapeats seas deposited these sandstones, during hundreds of millions of years while erosion removed several miles of rock from where we were standing, scouring away an ancient mountain range supposedly as imposing as the Himalayas.

Great also reflects the Great Unconformity's worldwide distribution. Rock even older than the Vishnu Schist downstream exists somewhere on all the continents—the Canadian Shield, Russia, South Africa, western Australia. But it seems that massive erosion scraped every continent clean of much of the younger rock deposited between two billion years ago and the beginning of the Cambrian, messing things up for paleontologists because those rocks held fossils they needed. This fooled early paleontologists into thinking the Cambrian was the time when life began. Cambrian rock was just the rock that was still there, and it contained fossils of multicelled creatures. Now we know that multicellular life boomed hundreds of millions of years earlier, during the marine radiation of Ediacarians, a bizarre radial fauna now extinct. Stromatolites of colonial cyanobacteria have been around for 3.5 billion years, and blue-green algae even longer. They gave us our oxygen atmosphere.

Here in Carbon Canyon this interface is so subtle between two sandstones that most people do not notice it. Yet with one step each of us "time warped" by crossing an interface representing the missing geological formations of a half billion years, one ninth of the planet's local history stolen by erosion.

The duration and magnitude of the erosion necessary to create the Great Unconformity seems almost impossible. One estimate is that erosion today skims one inch off North America every 850 years—100 feet in a million years. Before land plants like cryptogamic lichens evolved to hold the soil, erosion was so much faster that during the period of the Great Unconformity most Precambrian sedimentary rocks simply vanished as if the earth had been peeled like an onion. The 200-million-year housecleaning here ended only when the Tapeats Sea started depositing its coarse sands.

Our next steps carried us through a deep trench carved into weird and sinuous curves and sculptures that compel one to touch them. Most of my people did. Upper Carbon Canyon is such a narrow, shady, cool world unto itself that people expect it to narrow infinitesimally and vanish altogether near the top in the knife-edged crack of a cul-de-sac. What happens instead is stunning.

At the top the Tapeats Sandstone makes an abrupt ninety-degree turn upward from the horizontal, as if the rock were suddenly rubber. Here, truncated by the Butte Fault, the sandstone vanishes eerily, with its broken ends jutting into the sky like the fossilized skeleton of some gigantic prehistoric beast. The landscape beyond this break is suddenly wide open and desiccated by an unblinking sun. An hour from the river, we discover the size of Grand Canyon.

I glanced at everyone. For a moment they simply stood in the last patch of shade to stare across this vastness of the inner Canyon, their cameras forgotten. The panorama of gently rolling blue and purple hills of the Galeros Formation of the rare Chuar Group (a series of eroding, sloping siltstones, sandstones, and limestones visible nowhere from the river) stretched a half dozen miles under a blue sky to end at the familiar red rampart of the Paleozoic sequence—a 5,000-foot cliff whose base was Tapeats Sandstone and whose top was Kaibab Limestone of the Walhalla Plateau. The expanse of almost normal landscape between the familiar cliffs in the distance and us made me feel both microscopic and at home. This valley seems a lost world. One almost expects to see some strange prehistoric beast lumber across the low ridge.

Why this bizarre change in the landscape? To the north-northwest, the East Kaibab Monocline slices into Grand Canyon from the northwest and parallels the river southward for about twenty miles. Here, about halfway from its entry, it passes right beneath our feet. It has opened this region east of the 8,000-foot-high Walhalla Plateau to erosion that has carved a canyon suddenly six to ten miles wide: a true Grand Canyon and a land watered by perennial creeks (Nankoweap, Kwagunt, Chuar, Unkar) flanked by cropland, deer, and mesquite groves once vital to the Anasazi Indians.

Here at the base of the tortured and truncated Tapeats are stunted hedgehog cactus, prickly pears, creosote bushes, Mormon tea, and a few other desert shrubs populating a rarefied garden. Gusts of wind vibrated the creosote, brittlebush, and mesquite violently. A few yards farther into the sunlight and tumbled into the dry sand in the creek bottom (the trickle of alkaline water runs under these sands), was a weird purple-brown boulder four feet thick whose twisted columns squirm petrified within it.

"I suppose you're wondering why I brought you all here today," I started.

A few people rolled their eyes, but everyone was curious.

"Well, your perseverance has been rewarded, because we are in sight of a wondrous relic left by creatures who preceded us here in distant time."

"What's this thing called again?" Fabry asked me.

"A dinosaurian coprolite."

"A what?" he asked, knowing I was putting him on but unable to remember what it really was. By now everyone was curious.

"A fossilized dinosaur turd. And, from the size of it, probably from a Brachiosaurus."

"Come on."

"Okay, it's not a dinosaur turd," I admitted reluctantly. "It's a stromatolite."

"A what?"

"A stromatolite. A fossilized colony of one-celled photosynthetic bacteria. Hard to tell what formation it tumbled out of, but I'd guess that it came from the Galeros. A billion years old."

"That old?" Mary, one of our passengers, asked.

"Close. But that's not especially old for this fossil type. They come a lot older than that."

"How do they know that that's what it is?" she asked, eyeing it suspiciously. After the dinosaur turd, who could blame her?

"Good question," I said, "particularly because this is such a rare situation in paleontology. At first, people had no idea what these fossils were, or even if they were really fossils. Stratigraphically it was obvious that they were older than any other living thing that had left unmistakable fossils, but exactly what they were was a mystery.

"The eureka came when some invertebrate zoologist probed some strange bacteria living off the western coast of Australia in the shallow intertidal pools of a place called Shark Bay. These bacteria are colonial and photosynthetic. They secrete a jellylike ooze that traps silt in their colony but that allows the upper cells to live, so each colony grows up in the shape of a barrel. These barrels even orient themselves toward the sun. Anyway, these cyanobacteria were growing into stromatolites exactly like these mysterious fossils from three to three-and-a-half billion years ago. The bottom line is that this rock represents the oldest incontrovertible fossil type of living things on this planet."

"Not a dinosaur turd?" Fabry asked dryly, as if to point out that my calling it that had been dumb.

"Well, maybe," I shrugged. "Anyway, if anyone is wondering why the landscape made this abrupt transition, I'll try to explain. This break was caused by uplift along the western side of Butte Fault, a major and ancient fault. And then by erosion. That's the easy part. Wait, look at this."

I unfolded my large topographic and geological map of the Canyon, a work of art that impresses me every time I look at it, then pointed out the fault system here at Carbon.

"What's tricky is that all of this rock in front of us is billion-year-old Galeros Formation, four thousand feet thick. In fact, this, plus two thousand more feet of other formations in the Chuar Group, exist only here for a few miles along this west side

116

of the fault and nowhere else in Grand Canyon. Maybe nowhere else in the world. You'll see as we head downriver that this six-thousand-foot-thick series of rocks is missing from every place it should be beneath the Tapeats. Gone. Erosion took it all between six and eight hundred million years ago.

"So why is this stuff still here?" I asked rhetorically when no one obliged. "It must have been protected from erosion by being buried for hundreds of millions of years before the Tapeats formed. That could have happened only if this west side were downthrust deep into the earth. This means the crust on Butte Fault once moved down, and the east side moved up, before the Tapeats was formed; and that movement along this fault has reversed itself since then."

I paused. The blank looks on some faces now convinced me that I had become too arcane. I folded up the map. Dinosaur turds were more fun.

"Okay, everybody," I said, rising to my feet and shouldering my pack. "Time to move on. The traverse cuts over the Galeros into Lava Canyon." I gestured south. "If you still have water, drink it now. There'll be water in Chuar Creek." As people opened water bottles to gulp their precious fluid, I marveled again at our propensity to hoard—some of the hikers found dead of dehydration in the Canyon were carrying canteens partly filled with water.

Single file, we tramped parallel to the fault and up a low rise in the Galeros. Then we dropped into the head of a wash and descended along its dry bed. I could have pointed out the hidden face of Chuar Butte at our backs. At noon on June 30, 1956, "on a reasonably clear day" two airliners both flying east from Los Angeles somehow collided above Chuar Butte. Pieces of the demolished planes ricocheted off the cliffs and slopes and scattered down into tributary canyons and the river. All 128 people died. This was the worst peacetime aviation disaster to that time. Now most of the wreckage had been removed.

This route contrasted with everything else we had hiked in

the Canyon: flat, packed, coarse sand in a wash, like a trail in a city park. We strolled between purple, blue, and orange members of the Galeros and rubbernecked to gaze at the ramparts of black basaltic lava looming between the river and us. These were pillow basalts that erupted in a shallow ocean even before the Galeros was deposited. They loomed higher now, upthrust by the Butte Fault.

We halted at the brink of a fifty-foot waterfall that transformed our too-good-to-be-true trail into a suicide jump, then detoured around it via a steep traverse. We soon reached Chuar Creek and followed its gurgling course to the river between steep walls of bright rust in the Dox Sandstone.

The mouth of Lava Canyon (not Prospect Canyon, which forms the infamous Lava Falls) frames a view across the Colorado in Lava-Chuar Rapid (Mile 65), where the river races toward the Pacific. Here I explained that if anyone wanted their first warm bath instead of an arctic one in the river, this was the place—not *in* the creek, but well away from it in a bailing bucket filled with ninety-degree water from the creek. They cheered as if I had yanked a hidden case of champagne from the willows. The Colorado is that cold.

Sundown. The alpenglow on the Palisades of the Desert, the cliffs bordering this upper end of Furnace Flats to the east, had faded to a dull red, then to indistinct pink. Stars now winked over the gateway to this valley where the Anasazi had gained their greatest foothold in Grand Canyon. In fact, ruins lay buried in the sand a stone's throw from this camp. Now the beach was deserted. Moonlight washed the Palisades looming above our backs. There were ghosts ahead in this valley, and tomorrow we would meet them.

8

Enemy Ancestors

Gravel rattled under my sandals as I ascended the steep bank to Unkar Delta (Mile 72). It was only mid-morning but already the landscape was hot. By afternoon these acres would be an oven.

Taking the risk of spiking my toes with prickly pear spines, I glanced at the Paleozoic cliffs encircling us. Here, and in the seven miles upstream to Lava Canyon, these distant cliffs guard an inner valley of rounded hills we call Furnace Flats. Grand Canyon is ten miles wide here, and only by a few routes can we climb or descend the cliffs. Furnace Flats is a lost world, one stranger than fiction, a land where a civilization rose, thrived for centuries, then mysteriously vanished.

This solar-dehydrated Shangri-la owes a debt to the extreme erodability of the Dox Sandstone, a 3,000-foot-thick formation of the Grand Canyon Series whose greatest exposure along the river occurs here, between Miles 63 and 73. The Grand Canyon Series has suffered the same geological fate as the Chuar Group containing the Galeros—erosion has nearly erased it from the face of the earth. Only a few tiny islands of Grand Canyon Series survived—all within the Canyon—and the Anasazi settled on each of them.

At Furnace Flats, soft redbeds of Dox have eroded into deltas and hills—rare items in Grand Canyon—that could be farmed. Today rainfall here during the summer averages only three inches (and probably totals only seven or eight during the whole year), leaving the terrain too arid for farming. But between A.D. 900 and 1200, especially during the final century and a half, the climate here was wetter. During the wettest years, Anasazi had settled in the Canyon to wrest a living from the most isolated enclave in their hidden realm.

We guides who row trip after trip down the Colorado and consider Grand Canyon a second home are the first to admit that we could not survive here more than a few days without siphoning off food and equipment from the outside world. It would be hard not to respect and marvel at the ability of the Anasazi to survive here. It would be even harder not to wonder how they did it.

Where these mysterious people came from, why they came, who they were, and where they went after they vanished from here are only a few of the unanswered or partially answered questions surrounding the Anasazi. Even their name has nothing to do with them. *Anasazi* is Navajo for Enemy Ancestors. The Navajo and their Athapaskan cousins, the Apache, apparently arrived in New Mexico as early as 1300 from a long migration from western Canada. The arrival of Navajo raiders in northern Arizona followed the decline of Anasazi culture by five centuries. Navajos found thousands of abandoned stone pueblos scattered across the vast land they were claiming, and they called the vanished builders Anasazi because the pueblos were similar to those built by the Navajo's enemies: the Hopi south of Black Mesa and other surviving Pueblo IV people. What the Anasazi called themselves was a mystery—one that I now reckoned was solved.

Here on Unkar Delta the Enemy Ancestors left clusters of their most diverse settlements in Grand Canyon. Others are scattered upstream seven miles across Furnace Flats to Lava Canyon and beyond to Kwagunt, Nankoweap, and South Canyon—in the river corridor, up tributary canyons, and on rims and

120

mesas all the way to Soap Creek at Mile 11. Ruins lie scattered downstream too at Shinumo, Tapeats, Deer Creek, and beyond for another 170 miles. A map of the 2,700 known prehistoric sites in the Canyon (a list that is growing rapidly as structures emerge from the eroding sand or are discovered hidden up improbable side canyons) reveals that during their heyday in the eleventh century at least a thousand Anasazi lived in Grand Canyon, and maybe a lot more. Again, the 125 acres of field and slope in Unkar Delta is the best-known site of these Pueblo II people.

It is impossible to gain a real feeling for Grand Canyon without seeing what the Anasazi left behind. Indeed, now that I have seen so many of their hidden villages and granaries, their crumbling stonework half-buried in the blowing sand and slumping soil, their sherds of elegant pottery decorated with geometric designs that trigger aesthetic neurons, I can't stay away.

I hiked along the delta, a huge, nearly level ancient fan of soil and rock abutting the receding cliffs, then stopped at the dwellings perched on the edge of the north embankment overlooking the river above Unkar Rapid. Here I experienced another weird time-warp sensation, as if I could forge a rapport with this vanished past if only I tried hard enough. I suppose this experience also drives archaeologists to probe in the dust for long hours—or years. It now spurred me to imagine the scene were we to travel back in time and meet these tough little Indians dressed in brocaded cotton. We would seem like monsters: large, pale, clumsy, garishly clothed, with dark lenses hiding our eyes, and carrying strange black objects around our necks that could not possibly serve a useful function. In their world, in fact, *we* probably would serve no useful function.

I stepped over a pair of double-thick walls of limestone and sandstone mortared with clay and sand that had resisted a millennium of monsoonal poundings. The ruins faced a green river flowing against a bold cliff of maroon Dox on its far side, then plunging, roaring, into Unkar Rapid. This view, outside the door of a condominium today, would cost a million dollars.

The multiroomed ruin was labeled UN-9 by archaeologist Douglas Schwartz in the 1960s. I stared at the strangely angled walls of thick stone and at its storage cysts (walled storage pits) and firepits lined with thin slabs of Dox Sandstone. These had once held fires to cook hard-won food in corrugated gray pottery.

Nine hundred years ago, an Anasazi family lived here. Nearly a thousand years ago, communities of various branches of Anasazi lived throughout the Four Corners region of Utah, Colorado, New Mexico, and Arizona. Based on the pottery here, this family was Kayenta Anasazi—related to people from the high country east of the confluence of the Little Colorado and the Colorado in the Kayenta and Black Mesa area. They survived here by farming the classic triad of crops domesticated in Mesoamerica: corn, squash, and beans—three among a hundred crops domesticated by Indians. They also grew cotton for weaving cloth, but corn was vital, as it had been for 2,000 years in Arizona. They terraced the upland delta for dryland farming and diverted Unkar Creek to irrigate the lowlands. Doing both hedged their bets. Heavy rain might send a flash flood down the creek and wipe out their irrigated fields, but at the same time it would water their dryland crops. Hopi corn needs only two good rains to bear ears. On the other hand, if not enough rain fell, these Indians would have been forced, like the Hopi, to haul water to their dryland crops, while the irrigated crops would grow with less work.

Despite their sophisticated agriculture, life was no bowl of squash blossoms. This land was marginal and the climate unreliable. Their middens reveal that the people subsisted half on what they grew and half on what they gathered from the wild. The honey mesquite trees scattered thickly along the old high-water line probably staved off starvation many times. Probably so too did catclaw, fourwing saltbush, Indian rice grass, prickly pear, and agave.

At my feet, on the hard-packed earth in the center of the largest room in UN-9, sat a sandstone metate, a kitchen appliance abandoned a millennium ago but still usable. The Anasazi used metates to grind seeds and corn. The men here also hunted rabbits, bighorn sheep, deer, and other mammals.

This ruin is one of dozens here. Others are scattered in contemporary communities a mile upstream on both sides of the river at Cardenas Creek. This family must have hobnobbed with those people. No evidence for boats has shown up, but the Anasazi must have had some means of crossing the river other than waiting for winter to lower the flow enough to wade across. Rafts of driftwood?

I glanced across the barren, stony soil of the delta, dotted with sparse cacti and Mormon tea. It looks as if nothing edible would grow here. No one who has visited this delta has failed to ask, How could they have done it? The truth is that they could not do it today without diverting the river, which fluctuated so wildly prior to being dammed that it would have destroyed their diversion. This is why the Anasazi left. The climate of Unkar Delta has changed since they farmed here. National Park Service archaeologist Jan Balsom reckons that the Puebloans were here off and on between A.D. 900 and 1150, with possibly ten families living here during its heyday. She recently analyzed the pottery of the major occupation of UN-9 and dated it to between 1070 and 1180. The major lesson is this: the Indians apparently lived here only during years of a lot heavier rainfall. During the normal, drier times, they vanished. Finally, during the permanent drought after 1200, the Anasazi abandoned Grand Canyon and most of their other lands as well.

Archaeologists have reconstructed this history tidily—to the exact year—by dendrochronology, the science of interpreting tree rings. During warm seasons, temperate trees add a growth ring whose thickness depends on rainfall and the length of the growing season—they basically "photograph" their own growing seasons. By counting rings in core samples from old living trees, then cross-dating by overlapping them with those of older dead trees built into Anasazi structures, dendrochronologists have reconstructed the patterns of rainfall for every year in the Southwest back to 273 B.C. But because the wood used here at Unkar, while datable by dendrochronology and by another technique, carbon 14 analysis, was driftwood that may have been old even before the Anasazi dragged it up here for lumber, the earliest dates for these ruins came from the styles of pottery found buried here. The earliest might be A.D. 1000, but fifty years of high rainfall between 1040 and 1089 in the Southwest corresponded to the biggest explosion of Anasazi culture across the Four Corners area.

By Canyon Anasazi standards, UN-9 was a palace, a complex of double-thick and triple-thick walls containing twenty rooms, cysts, and ramadalike enclosures. It held five firepits. But the walls were made of stone probably only halfway to the roof beams. From there up they were wood and mortar, called jacal. Intact roofs here in Grand Canyon are beamed and thatched, and mortared with adobelike mud. Only two of the rooms opened into the central plaza at floor level. Apparently the rest could be accessed only through roof entries. During UN-9's century or more of use and three or four occupations, the Indians remodeled some of its rooms; one large room was remodeled twice. Some rooms also had three superimposed floors.

What were the Anasazi like? They were at most five to five and a half feet tall. Their worst hazard was worn-out teeth. Grinding their food on sandstone metates left grit that abraded their teeth. If you were an Anasazi, you could expect your teeth to wear out, and chewing is so vital to the assimilation of food that worn-out teeth probably led to poor nutrition and a shorter life.

I turned away from UN-9 to walk up the gentle slope toward the red terraces of Dox Sandstone. My nose kept pointing toward the ground; I could not help searching for artifacts abandoned nine centuries ago—fourth-dimensional telegrams from the Enemy Ancestors to us.

The vast beauty of this place protected by mile-high cliffs camouflages its teeth. Even during wet times, this had been a harsh place to eke out a living, and contrary to what many people guess, the Anasazi probably did not migrate into Grand Canyon simply because they suddenly could. As Douglas Schwartz and his colleagues suggest, they likely were forced to farm here.

Why? Because for the Anasazi, more rain in the Canyon was a good-news/bad-news proposition. It did not simply make agriculture feasible in lower elevations; the additional rain was due to a generally cooler period that would have shortened by a few days the growing season on the Kaibab and Walhalla plateaus. Today the growing season on the *lower* portions of the Walhalla Plateau is almost exactly the 104 to 105 frost-free days required by some Indian corn to reach maturity—with no margin, although this is plenty long for beans. Schwartz suggests that the Anasazi invaded Unkar Delta during these wetter periods when a shorter growing season on high elevations along the rim suddenly made it impossible to grow corn there. They monitored the celestial cycle, identified the solstices and equinoxes, and knew exactly when to plant. Over the years the Anasazi were forced down into the Canyon by moisture and cooling, then up again by drought, seesawing with the oscillations of climate inexplicable today.

But this forced yo-yoing of Indians up and down the Canyon did not dictate shifting their *permanent* habitations from rim to river and back. Archaeologists Helen Fairley and Jan Balsom suspect that families may have kept two dwellings—one on the rim to grow beans, which required wetter land, and another in the Canyon to grow maize and cotton—with members at each to tend the fields. Balsom analyzed the mineralogy of pottery from Unkar and the Walhalla Plateau of the North Rim (inhabited simultaneously with Unkar) to test this idea. The ceramics reveal that Balsom was right about two-house families: pottery was not only of identical styles, it was also made from the same clays mixed together from both areas.

A few hundred yards upslope from UN-9, at UN-4 and UN-6, the architecture differed. Instead of building storage bins, cysts, and firepits inside the dwellings, the families there had built them outside. Why the change? Maybe it was just a fad. Or perhaps the people at UN-9 built when predatory Utes raided them and forced them to conceal their hard-won food, while the people upslope did not. Anasazi throughout the Canyon built their granaries—some as small as televisions, some as large as refrigerators—in alcoves high in cliffs. Because these granaries were camouflaged, far removed from dwellings, and sometimes risky to reach even when one knew where they were, it appears they were safe-deposit boxes for surplus corn and beans. But because these niches in cliffs also offer the best protection for grain against rain and rodents, granaries alone cannot tell us that these Anasazi suffered from raiders.

A clue that they were indeed subject to raiding exists across the river and a few hundred yards upstream, high atop a hill overlooking a settlement at the mouth of Cardenas Creek and the settlements here and a mile upstream dating from around 1150. It is a large, well-built stone structure. No midden of trash or broken artifacts is nearby. It commands a view extending 360 degrees and allows surveillance of every route penetrating this region. A few miles upstream I visited another "tower" built on the high ground of Furnace Flats that provided a panoramic view, allowing the detection of invaders in plenty of time to set up a defense. Were these "towers" built for this reason—or instead as male society rooms. Or both? Weighed with all the other quirks of Anasazi architecture—the granaries hidden high in cliffs and entire villages hidden up tiny, dry side canyons where hauling water would have been a constant chore, entries for dwellings only though their roofs, and the fact that several multiroom dwellings here had been burned (intentionally?) and then later rebuilt—convince me that at least some Anasazi were preyed upon and hence were building defensively. But against whom? Utes?

Higher up the gentle slope I stepped over long parallel rows of boulders the Indians had hauled here to terrace the delta and conserve the soil into small fields that retained moisture. Once, while taking passengers on a grand tour here, I spotted an arrowhead flaked from gray chert on this stony soil. My eyes refused to move on. Another telegram from the fourth dimension. I had seen several arrowheads on this delta before, but not this one. It was a lot larger than a typical Anasazi point. I debated whether I should hand this projectile point around to everyone, risking that it would vanish. I decided to keep moving. Then I wondered, Was it an Anasazi projectile point, or had it once been hafted to the arrow of an enemy raider?

While beautiful, many Anasazi projectile points are so minute that it taxes one's imagination to understand how any craftsman could control his pressure flaking of stone to such a fine degree. Douglas Schwartz's team sorted through almost 10,000 stone artifacts, but only 98 of them (mostly of chert) were bifacial—the type of well-shaped projectile point or knife that most people think of. An arrowhead must be light, aerodynamically true, and sharp. Anasazi points meet these criteria beautifully but still make me wonder why they were so small.

Finally I arrived upslope at UN-4 and UN-6. Here walls in the same dwelling differed from one another in both construction and type of rock. Some dwellings had been occupied, abandoned, and rebuilt three or four times over a century. I stared down at a broken metate cemented into part of a foundation. Does this explain why so many other walls on this delta are so short? Did late arrivers find the best stones already taken and do what anyone else would do—cannibalize abandoned dwellings for easy building materials?

It seems unfair. The Anasazi vanished despite an ingenious and superior technology that allowed them to wrest an elegant life from a land so forbidding that, when people look at it today, they can only shake their heads.

After the last climatological respite between 1190 and 1209, permanent drought drove the last Anasazi from Unkar and the rest of Grand Canyon outside Havasu Canyon. This drought coincided with the death of many other communities of Anasazi in the Four Corners region, including the gigantic Chaco Canyon complex (900–1115) of New Mexico in the San Juan drainage. Within Chaco Canyon, thousands of Anasazi lived in nine "great houses," such as the five-story Pueblo Bonito, which contained 650 rooms. Chaco Canyon was the social center of a region of roughly 50,000 to 100,000 square miles dotted by twenty other great houses interconnected by well-constructed roads running as straight as arrows. While the extent and chronology of the climatological disaster here in Grand Canyon are unknown because most of the ruins here haven't been dated, the disappearance even of *all* the Anasazi here was a drop in the bucket compared to the huge list of casualties among their contemporaries in the greater Southwest, where an entire vast, complex, and interconnected civilization nearly vanished.

Where did the Canyon Anasazi go? On spaceships to another planet? Probably not, but no Canyon Anasazi mystery is more compelling. Schwartz and his colleagues speculate that the final drought caused a "general contraction of population back toward the Kayenta heartland." But this meager "heartland" was probably filled already, so it could not have accepted all the refugees. Many must have been forced to search elsewhere for new homes. But where?

The century and a half after 1150 marks the Pueblo III, or Classic Anasazi, period (Pueblo I spanned the years 700 to 900, which followed 1,700 years of Basketmaker maize farmers). The consensus is that Anasazi from the north and west, including those in the Canyon, emigrated to three major destinations: south to central Arizona, southeast to west central New Mexico (Zuni country), and east to northern New Mexico along the upper Rio Grande. But this too is an oversimplification. Between 1150 and 1250 the Anasazi elsewhere exploded in population and built by far their most impressive stone cities, housing more than a thousand people. Some of these far-flung cities—Wupatki, Canyon de Chelly, Hovenweep, Mesa Verde, and others—

became cultural centers whose complexity nearly equaled those developed in Mexico and farther south, and indeed probably owed a cultural debt to them.

As the Anasazi abandoned these cities in turn due to climatic change and habitat deterioration from overexploiting their soils and forests, which in turn led to their overexploiting wild animals and plants, one group after another migrated again and vanished archaeologically. Because so many abandonments seemed due to habitat deterioration instead of drought, paleoecologist Richard Hevly concludes of these Puebloans, "Man is the dandelion of the animal world, blowing hither and yon, settling on any appropriate region of the landscape, and then destroying the very qualities that were originally sought."

Some Anasazi seem to have merged with Puebloan Mogollon people inhabiting the mountains along the southern part of the present-day border between Arizona and New Mexico. Others merged with the Puebloan Salado (Spanish for "salt") culture farther west between the Mogollon and Hohokam. Closer to home, the Sinagua (Spanish for "without water") people who settled east and north of Flagstaff seem a mix of Anasazi, Mogollon, and Hohokam. The Sinagua suddenly became Anasazi-like. Why?

I stared at the South Rim of the Canyon. What lay beyond was hidden, but there, beyond, among the Sinaguans, is where the big clues lie. The Sinagua blossomed in the regions of Walnut Canyon and Wupatki east and north of Flagstaff a century after Sunset Crater erupted most violently in 1064 (during the next 186 years of sporadic eruptions, Sunset Crater blasted more than half a billion tons of fertile ash over 800 square miles east of the San Francisco Peaks). Dendrochronology also reveals a trend of warming and increased summer rains between 1050 and 1150. With intriguing timing, the Sinagua culture boomed all over the San Francisco Peaks region, including Wupatki, Walnut Canyon, Three Courts Pueblo, and hundreds of other settlements as a blend of Sinagua, Kayenta Anasazi, Winslow Anasazi, Cohonina, and Prescott cultures and reached its greatest complexity around 1150. This was when the Kayenta Anasazi from Unkar Delta and the rest of Grand Canyon made their final exodus. Hardly a co-incidence. Only about fifty miles separate Unkar from Wupatki (Hopi for "Tall House"), and only another twenty miles lie be-tween Wupatki and Walnut Canyon.

With nearly a hundred rooms, Wupatki is the largest and most technically impressive of hundreds of settlements among the 2,700 sites of the Sinagua and Pueblo III Anasazi cultures around Wupatki National Monument. Were the Unkar Anasazi part of this Sinaguan line? It seems likely. But by 1190 building had nearly ceased, and in 1215 another drought began. By 1225, aridity and a loss of soil fertility (some prehistoric cornfields at Wupatki are still depleted of nitrogen after more than 700 years) and a loss of moisture-retaining ash had reduced the Sinaguan boom near Sunset Crater to a bust. By 1300, Wupatki held only ghost pueblos.

Although this bust too was tied to environmental deteriora-tion, it was not that simple. A burial found at Walnut Canyon held a woman with a non-Sinaguan arrowhead lodged in her chest. More sinister evidence was unearthed by archaeologist Watson Smith and analyzed by Christy Turner near the northwestern edge of the Wupatki region in Big Hawk Valley at the House of Tragedy and at Three Courts Pueblo, where a mix of Anasazi, Cohonina, and Sinaguan peoples had lived together. Smith's team excavated several rooms, even a sacred kiva, where each turn of the trowel unearthed evidence of violent death. Corpses of both adults and children—and even infants—had been dis-membered by humans and scattered helter-skelter. Many of the long bones had been cracked open as if to extract the marrow. No one ever returned to bury these dead—burying was normal among these Puebloans—nor were these dwellings inhabited again.

Were these murders committed by cannibalistic raiders? Had similar carnage occurred at other settlements, to be cleaned up later by survivors who moved back in, such as here at Unkar? Archaeologist David Wilcox investigated the surge of defensive fortifications among the contemporary Hohokam farther south and concluded that it was a reaction to warfare. The Sinaguans also built many defensive "forts" atop hills. The role of violence in the Anasazi migrations may remain a mystery, but this is certain: Not all abandonments in Anasazi country were merely a reflection of industrious people fleeing drought for more fertile fields.

Also, not all the Puebloans in northeastern Arizona vanished. The survivors, the Hopi, recount a tribal exodus from Grand Canyon. When life became untenable in their Third Underworld due to harsh environmental changes and social disharmony, their ancestors abandoned it to emerge from the Sipapu into the Fourth World. The Sipapu is a sacred site, a massive travertine spring that looks like the top half of a giant pumpkin fifty feet across and half buried in the Tapeats Sandstone next to the Little Colorado River four or five miles upstream from its confluence with the Colorado. About six feet down in the center of the Sipapu, clear water bubbles forth to fill a hidden pool. Despite my doctorate in biological ecology, this spring half convinces me that it *could* be a portal between our world and some other. Jammed in cracks in the travertine above the bubbling spring are prayer feathers. Hopi mythology holds that it is through this portal that the dead return to the world below. More important, Hopi mythology holds that their ancestors *emerged* here, a one-day, fifteen-mile walk from Unkar Delta.

Archaeology yields more telltale connections. Prehistoric Hopi differed little from the Anasazi at Unkar and the Sinaguans east of the San Francisco Peaks. Further, during the century beginning around 1250—half a century after Unkar became a dust-blown ghost pueblo and when Wupatki was being abandoned—the population sixty miles away at the Hopi Mesas exploded.

More telling, the Hopi today still recognize Wupatki and many other Sinaguan pueblos (Awatovi, Homolovi, and others) as ancestral villages and still revere the San Francisco Peaks west of Wupatki as the home of the sacred Katsinam. At Homolovi around 1100 the black on white pottery was Anasazi. During the next 150 years it evolved into black on orange, black on yellow, black on red, and polychrome—in short, Anasazi pottery evolved recognizably into prehistoric Hopi pottery.

Clearly, some Sinaguan–Western Anasazi evolved to become Hopi. But John McGregor's excavation near Walnut Canyon of the most elaborate Sinaguan burial yet discovered, the Magician's Burial, clinches this connection. It held 600 burial offerings: 25 decorated pots, 450 arrowheads, beautifully inlaid pieces of jewelry, and 12 wooden wands carved and painted to resemble hands, hooves, and other forms. On a hunch, McGregor questioned several Hopi about these finds. Surprisingly, each Hopi not only knew the function of each carved wand but identified the ceremonial society to which the dead Sinaguan had belonged 700 years earlier as the Motswimi, or Hopi Warrior Society.

Another eerie connection became apparent when a Hopi excavator at Wupatki was asked by a Hopi of another clan whether he had ever found a man buried with a dog and parrots. "The question surprised him because they had just excavated such a burial," Scott Thybony writes. "The curious Hopi said members of his clan still remember the buried man's name. He was a chief called Chipiya."

The mystery of what the Anasazi called themselves is clearly no longer really a mystery. The Hopi call their own ancestors Hisatsinam. More to the point, they call themselves Hopitu Shinumu, the People of Peace.

Why did the Anasazi–Hisatsinam–Hopitu Shinumu disappear everywhere else if their technology was so advanced? The reasons offer us a vital lesson in survival. "The nature of the abandonment of so large an area as well as the causes for it," archaeologist Fred Plog notes, "remain one of the great unexplained events in Western Anasazi prehistory." But death, migration, and aggregation into ever larger communities in the best remaining habitats shrank their territory and placed impossible demands on the land. "Erosion, crop failure, disease, and droughts occurred in varying degrees in different areas and magnified the problem until most of the Western Anasazi area was empty," Plog concludes. "Remaining peoples concentrated in the Zuni and Hopi areas."

How did the Hopi prevail? "The Hopi descendents of the Western Anasazi," Plog writes, "survived using a risk-hedging strategy: planting crops in small plots and in a variety of environmentally different situations spread out over hundreds of square miles." Plog also notes that "they planted enough in any growing season to last for three. Had the Spanish not interrupted developments in the area, most of the Colorado Plateau would probably have been repopulated by peoples using a strategy such as this one."

Saying the Spanish "interrupted developments in the area" is an academic euphemism for war, rapine, and genocide. Coronado's invasion of western North America, a spin-off from the rivers of gold flowing into Spain from the conquests of Mexico and Peru, was a Puebloan death knell. It occurred by the thinnest of threads—the tales of a black slave named Esteban shipwrecked in 1528 on the coast of Texas—but it violently shattered Puebloan culture. In Mexico City, Esteban reported that Indians in Texas had claimed that cities existed along the upper Rio Grande that had been built by farmers wealthy with pearls, gems, copper, silver, and gold. In 1539, Fray Marcos de Niza trekked north in search of these Seven Cities of Cibola. He took Esteban and a company of Indians with him.

De Niza ordered Esteban and a few Indians 150 miles ahead to Hawikuh (in northern New Mexico). Hawikuh, the first of the seven Zuni Pueblos, was a four-story, 200-room "city" built of mortared stone and log beams that walled a plaza controlled by a single gate. The Zunis so resented Esteban's arrogance that they killed him on the spot (Fred Kabotie, a Hopi artist, writes that the Cha-Kwaina Katsina still seen in Hopi and Zuni ceremonies represents Esteban). After de Niza's Indians rejoined him, he returned to Mexico City and told the viceroy that the pueblos were cities walled with gold. This being only twenty years after Cortez and his few hundred troops had toppled the Aztec empire and milked it of gold surpassing their most avaricious dreams, de Niza's tale was believed.

So, months later, on February 23, 1540, Francisco Vásquez de Coronado, twenty-six years old, led north a contingent of 336 soldiers, 1,000 horses and 600 mules, thousands of cattle and sheep, about 700 Indian allies, and six Franciscan friars (including Fray Marcos de Niza), who were along, as Marc Simmons reports, to see that "the conquest might be Christian and apostolic and not a butchery."

After four months of following Indian trading paths, Coronado's expedition located Hawikuh, attacked the Zunis, defeated them, and ransacked their pueblo. This was the first European military action in what became the United States. The conquistadores found no gold, silver, or gems, but the hungry soldiers reckoned the Indians' three-year stores of beans and maize, plus domestic turkeys, to be nearly of equal value.

Coronado pushed his army north, levying food taxes on all the pueblos they found and waging merciless war against the Puebloans who could not feed them. Learning of another seven cities in the northwest, at Tusayan—the Hopi towns—Coronado sent Pedro de Tovar to reconnoiter. A month later Tovar reported that the Hopi were no richer than the Zuni but that a great river existed to the west.

Coronado ordered Lieutenant García López de Cárdenas to find it. Cárdenas enlisted Hopi guides and traveled twenty days west of Tusayan to the rim of the Grand Canyon and "discovered" it. Cárdenas judged the Canyon to be eight to ten miles across from his vantage point on a forested plateau where the view to the north was still of desert, not forest. This narrows their location to the South Rim between Desert View and Moran

Point, overlooking Furnace Flats and the river here. They reckoned the Colorado River to be six feet wide, although the Hopis insisted that this was wrong. The Spanish found it impossible to descend.

Failure was serious for an expedition this expensive. Having followed every lead without confiscating a nugget of gold, Coronado set up winter quarters among the dozen pueblos of the Southern Tewa along the Rio Grande and extorted from them an impossible tax of food and cotton blankets. When soldiers molested Tewa women, two pueblos resisted. After a bloody siege the conquistadores won, taking 200 prisoners, whom Cárdenas ordered to be burned at the stake. Those not burned were cut down trying to escape. All 200 died. Tewa from nearby pueblos abandoned their homes and fled into the mountain snows.

In spring Coronado pushed his men into Kansas in search of a golden civilization called Quivira. By mid-1541, after having found nothing but ever poorer Indians on the plains, Coronado returned to Mexico City, where he was judged a failure—he had spent two years and two fortunes on a 4,000-mile trek that had turned up nothing but penniless Indians. Coronado was put on trial but was found innocent. Afterward he was crippled by brain damage suffered while racing his horse. Cárdenas was tried in Spain for burning innocent Indians and found guilty. He spent a few years in jail and forfeited part of his inheritance.

Sixty years later the Spanish returned to colonize New Mexico and convert the Puebloans by the cross and the sword—mostly the sword. They annihilated Acoma Pueblo, for example, which vies with the Hopi Mesas as the longest continually inhabited city in the United States—at least 1,000 years. In 1599, despite its natural fortification atop 400-foot cliffs, Vicente de Zaldívar and seventy soldiers forced Acoma to surrender and then killed 800 of its 6,000 people, hacked one foot off every surviving male over 25 years of age, and condemned everyone to twenty years of slavery. (Several Acomas eventually escaped slavery, returned, and rebuilt the pueblo.)

Eighty years later, on August 10, 1680, a warrior named Popé masterminded a war of independence fought by a Puebloan alliance that included the Hopi. They killed twenty-one of thirty-three missionaries plus 400 other colonists in New Mexico and expelled every other Spaniard back to El Paso. Though in 1692 the Spanish surged back into New Mexico by dividing and conquering in an occupation from which most Pueblos have still not recovered, the Hopi continued their firm resistance to the Spaniards and their religion, abandoning their towns near the springs at the feet of their three mesas to build defensible towns atop them.

Three Franciscan priests had converted some Awatovi Hopis in 1629, but after the Puebloan Revolt (during which all five priests in Hopi country were killed), the other Hopis warned Awatovi to abandon Christianity. They did not, and in 1692 only Awatovi did not fight the reinvading Spanish. Finally, in 1700, Hopis from the other towns raided Awatovi, killed all its men, and took the women and children captive. The governor of Santa Fe sent an army to punish them, but the Hopi repulsed the Spaniards, driving them back into New Mexico. In 1716 a larger Spanish force attacked Walpi (now situated atop First Mesa), but again the People of Peace defeated them. The Hopi later spurned Friar Garcés (in 1776) and Friar Escalante (who in 1775 insisted that a military expedition was the only solution). The bravery of these descendants of the Anasazi is all that saved them. Most tribes have perished utterly under the diseases and weapons of foreign invasions, but today the 9,000 members of the Hopi Tribe survive with their culture and land intact.

Today the Navajo and the Anglo invaders possess nearly all the rest of the Four Corners area that once belonged to the Anasazi. Here on Unkar Delta I stared at their crumbling foundations and wondered about the future. When our civilization has gone the way of those who rape the land beyond recovery, who will be the last to persevere here in these unforgiving deserts of northern Arizona? Despite being under the heels of the Anglos and being trapped in an enclave within the vast Navajo Reservation, all evidence indicates that the ultimate survivors will be the people who for at least two millennia have met every test of the harsh Southwest successfully—the First Americans who attained the highest levels of artistic and social development in North America and the most successful agriculture: the Hopi.

Possibly the next renascence of the Enemy Ancestors is yet to come.

9

Cliff and the Great Unknown

A drenaline vibrated my hands as I propelled us away from shore and across the eddy. We had just scouted Hance, at Mile 77 the first technically difficult rapid on the river. Scouting made choosing a route a lot easier, but it had taken only a glance from the steep sand dune at the mouth of Red Canyon to size up the danger. The rapid was gigantic and the waves were monstrous. The big holes out there could swallow elephants without a belch. The Colorado was flowing at 40,000 cfs. And here it dropped thirty feet, the biggest drop on the river. Besides that, it was June 6, 1985, and I had been watching for a helicopter to drop out of the blue sky in search of me ever since we had put in at Lee's Ferry.

I glanced downstream. At this flow only a left-to-right entry made sense. My game plan hinged on identifying the most benign "track" in the current and then finessing an entry to merge the boat with it. It also hinged on eliminating all lateral momentum that might cause the boat to deviate from the track and slide into a hole, then another and another. And even though the entire rapid was a football field wide, this left route was a narrow slalom through a mine field of potential disasters, mainly water pounding over submerged boulders to create holes that could flip or swamp my boat. Conveniently, the current line zigzagging between these snags to success could not have been more obvious had it been painted with a white stripe. All I had to do was enter in the right place. And then stay on track.

I glanced across the wide pool of emerald water above Hance to the right shore. The cliff beyond the pool held bright red beds of Hakatai Shale, one of the oldest formations of the Grand Canyon, or Unkar, Series. The colors were so rich that the scene seemed a technicolor trick. As if an antidote to this alleged illusion, a dike of dark gray diabase slashed diagonally upward through the redbeds from its origin in the rapid itself. This dike topped out against the Shinumo Quartzite atop the Hakatai. Rather than putting a damper on this spectacle, the dike of once-molten diabase was what caused Hance to be such a serious rapid.

Hance Rapid was created by flash flood debris flows from Red Canyon's habit of dumping trainloads of boulders into the river. But whereas the Colorado normally scours many boulders downstream of rapids to iron out most of them, here at Mile 76.5 the river collided with a complication.

A billion years ago, after the entire two-mile thickness of the Grand Canyon Series (including the Dox, Shinumo, Hakatai, Bass, and Hotauta formations) had been deposited, a county-sized dome of molten basaltic diabase squirted upward from deep in the crust. The diabase rifted up through the Hotauta Conglomerate but stopped partway through the Bass Limestone and spread laterally for hundreds of square miles. Like an immense hydraulic jack, it split the Bass horizontally and pumped open a huge, pancakelike gap in its middle, where the diabase accumulated in a horizontal sill hundreds of feet thick and an unknown number of miles wide. A billion years later, Red Canyon, along with the rest of Grand Canyon, was created by uplift and erosion to debouch exactly at one of the rare places where this diabase did breach the Bass and Hakatai. The complication here is due to the greater resistance to erosion of the diabase compared to the soft Hakatai Shale around it. This difference allowed the river to cut a mini Niagara Falls here as it rushed over and around the boulders from Red Canyon perched on the diabase, then suddenly poured onto the shale to erode it like sugar. This differential erosion made Hance Rapid the biggest in Grand Canyon—a 30-foot drop.

I shipped my oars as we crossed the eddy fence. The current feeding the rapid grabbed us. My route cut immediately left of the largest boulder out here, now half a minute away and nearly submerged in the high flow. I knew the current would take us there exactly. Many boatmen become so frantic during this wait that they row back and forth, overadjusting and second-guessing and taking twenty strokes when three would do, like a neon sign advertising lost wits. Instead, I now sat with oar blades dripping and watched as the current carried us. Cool.

But I wasn't cool. Today was the one day of this two-week trip when I could phone home, and I suspected that I would be hiking out immediately after that phone call—assuming that our area manager had found someone to row my boat and passengers downstream from Phantom Ranch. That touch-tone phone now seemed the instrument of destiny. Or was I already too late?

"Okay," I said to Karen, Chris, and Melanie, now packed hip to hip on the front deck. "Make sure you've got your life jacket fastened snugly. I'm going to enter this stern first, then pivot once we pass that big rock. The rest will be bow first."

They looked down at their life jacket clasps. Behind me, my fourteen-year-old son Mike checked his grip on the straps. This was old hat for him; he had been coming along with me since he was six and rowing since he was seven.

That huge rock suddenly accelerated toward us. I pulled two strokes downstream to slide my left tube into the powerful eddy behind that rock where the countercurrent would grab it and pivot the boat automatically. The eddy should also swing the boat—like a planetary probe such as *Voyager* being slung toward

Saturn by the gravitational pull of Jupiter—toward the right, where we needed to go.

It worked. I pulled on the right oar to enhance the pivot, then on both to increase momentum to the right, then yanked on the right again to complete the pivot. I checked my markers. My heart constricted. We had to be on. I would have only two or three seconds to fix a miscalculation. My approach had been elegant, but once you slop off the track at very high water like this, bad things can happen. And they usually will.

But we were precisely on track to thread between the next two pour-over boulders. Now all systems were go.

"Okay, hold on!" I yelled, thinking that people had to be crazy—or very trusting—to sit in the front of a boat in a place like this. "It's gonna get big, then a *lot* bigger further down!"

The boat buckled and bucked in a sharp hole, emerged, slammed and bucked in a second one that buried us in white water and then ejected us into the middle of the rapid, where the pounding current accelerated us into the land of the giants. Now we rode up and over mountainous roller coasters of pulsing water fifteen feet high. Big enough to be worth the price of admission. Big enough to make me wonder what I was doing out here.

The waves continued as if they would roll into infinity. Big, really big, huge, really big again, just plain big, not so big . . . Then it was time to pat myself on the back.

"Bail!" I yelled. Nicely. My passengers in front were women. Nice women.

Involuntarily, I glanced toward the sky again. No helicopter, but our bright yellow Domar riverboats would be easy to spot from the air.

I pivoted sharply in the tail waves to aim the stern downstream and right. For the first time I rowed hard instead of just smart, to break us out of the main current and into the first micro-eddy. This eddy offered us a ringside seat to watch Michael Boyle, Michael Fabry, Craig Alexander, and our trainee, an ex–

Vietnam fighter jock named George Franklin Keene III as they roller-coastered through Hance. If anything went wrong out there, we were in a perfect position to rescue swimmers before they were swept into lower Hance.

"Nothing had better go wrong," I whispered. I had a date with Ma Bell.

"That was great!" Chris bubbled. "You did a perfect job."

"Thanks," I said, fighting the urge to blush and say, "Oh shucks, ma'am, 'twern't nuthin'," just for effect.

Instead I watched intently as our next four boats roared through Hance. Once they had all sped past us into lower Hance, I pulled hard to escape our eddy and said between gritted teeth, "This rapid's namesake is downstream a quarter mile and up on that right slope. See those light cream-colored tailings just below that solitary catclaw? They're tailings from John T. Hance's mine."

A century earlier, Hance had been one of a few dozen miners trying to wrest a living from Grand Canyon. Unlike the others, he had struck deposits of long asbestos fibers here in the Bass Limestone. Serious pay dirt.

"Hance was a mule skinner in his forties on a prospecting trip when he first saw the Canyon in 1883," I explained. "Asbestos was in high demand then for weaving fire-proof cloth—for theater curtains. Good fibers were rare. The only reason they exist here is because the Bass Limestone contains a greasy green mineral called serpentine that, when that molten diabase fried the Bass, was metamorphosed into asbestos.

"But to get his ore to market," I continued, "Hance had to build two trails down Red Canyon from the South Rim. I've hiked the second; it's a great trail. He also built a cable ferry to get his ore across the river—down there where Boyle's boat is now," I pointed. "To load it on his mules.

"But the spin-offs were worth more than the ore. Tourists hounded Hance to guide them down here. At the time, no other reasonable trail existed into the eastern half of the Canyon. Eventually Hance started advertising in Flagstaff as a guide, and his business grew so big that he built a tent hotel near his log cabin on the rim. He became the first permanent white resident of the South Rim. He spent the rest of his life here."

As we zipped past the foot of Hance Rapid, we were suddenly floating on a narrow, impatient river rushing through a thousand-foot deep inner gorge hemmed in by tall, novel, jagged cliffs of vertically foliated dark gray schists and pink gneisses and granites. Here the Colorado veers from its long southerly course and bends to the west through the next 200 miles of Grand Canyon. This narrow inner canyon now hides everything beyond it and immediately becomes the only reality. Again the Canyon seems too big to be real. It feels as if we are at the bottom of the Earth. Four people have even asked me independently, "Are we below sea level here?" (not imagining how the Colorado River would drag itself back up to sea level to empty into the Gulf of California).

First Granite Gorge, this section of river, is impressive. It's also a dead ringer for the River Styx. Deep and final. Even the river itself is deep—up to 110 feet. And you can't turn back. Hance Rapid is the gateway to Adrenaline Alley: forty-odd miles of hard rock jutting from the river in ancient convolutions and tortured warps bearing testimony to serious subterranean abuse. The rapids here are gigantic and the rock old.

Older rock exists—in western Greenland and Minnesota, around Hudson Bay, in South Africa, in Zimbabwe, in western Australia, and in the USSR. But the rock surrounding us—this gnarled and brittle Vishnu Schist riddled with epileptic dikes of Zoroaster Granite—was about two billion years old, 40 percent as old as the planet itself. And it looked it.

The contrast with the orderly horizontal beds of Tapeats Sandstone now sitting several hundred feet above us is striking. This dark, vertically foliated schist, shot with veins of red granite and white quartz that have been twisted, squeezed, warped, and metamorphosed by the rough treatment of being slammed between two proterozoic microcontinents for 100 million years seems to encompass all history since the beginning of time. The metamorphosis of common mudstones to these vertically foliated schists required both being heated to sustained temperatures of about 1,200° F at a depth of about a dozen miles and being squeezed horizontally between the mammoth jaws of the vise of colliding continents for a *long* time. Even more amazing, this bedrock had been metamorphosed like this at least five times.

Everyone gazed silently at the jagged walls as if memorizing them. Every once in a while, Mike, Karen, or Chris would glance upward for that phantom helicopter. They knew what I was going through.

By passing Red Canyon at Hance we had committed ourselves to John Wesley Powell's Great Unknown. Although beautiful, it is not a friendly place. The rock is so resistant to erosion that most of First Granite Gorge is narrow, so narrow that campsites are nearly nonexistent, and the rock is so hard that tributary canyons are shorter and steeper. Because of their steepness, when they have flash flooded monstrously in the past, they have dumped bigger boulders than the ones scattered in the rapids behind us. Hence the rapids in First Granite Gorge are the biggest in Grand Canyon (with the exception of Lava Falls down at Mile 179.5 and the now submerged Separation and Lava Cliff rapids under Lake Mead).

We ran Sockdolager Rapid, then Grapevine Rapid. Each was mountainous, huge, scary, exhilarating. They needed superlatives. My Domar performed better than any boatman could ask for. But once through these rapids I became really nervous. Even though I had seen no helicopter for the past five days, I was worried about Connie.

For the eighteen days before our put-in at Lee's Ferry, I had stayed home with her. She was nine months pregnant. She was due. But day after day in our tiny apartment in Flagstaff, Connie had felt not the slightest twinge to hint that the event we were waiting for was imminent. Imminent, hell; it was overdue. I had canceled my trip in May to make sure that I would be with her when her time came. But it would not come.

Our situation was complicated. Less than two years earlier, Connie (and I) had been pregnant. But in her fifth month she had had complications, complications that her doctor had not handled conservatively. Connie miscarried on Christmas Eve, a perfect little baby boy who weighed one pound and lived less than an hour.

To bring her out of her sadness and to distance her from our loss, that spring I took us to Papua New Guinea to organize and market a new Sobek rafting operation to run trips on the Waghi River in the highlands. She rowed the oar boat and I captained the paddle boat. We traveled over vast parts of that huge, wild, and beautiful island, drumming up business and going native. As a finale we entered the Bulolo region, a territory terrorized not so long ago by a ferocious tribe of cannibals called the Kukukuku. From Bulolo we ran a six-day trip down the isolated wilderness of the Watut River, where she was faced with the most difficult white water she—or most guides—have ever seen from a rowing seat. It had been wild.

In the meantime Connie had landed a position as a nurse at the Grand Canyon Clinic. She returned, literally from our take-out from the Watut, to start work there while I consolidated our operation, then followed to join her a month later. As I ran the Colorado River that summer, Connie timed her days off to hike down Hance Trail or the Kaibab or Bright Angel Trail to meet me for a day. Two if we were lucky. The object: get pregnant. By the end of summer we had done it again.

But again Connie developed pregnancy complications, this time in her seventh month. Her new doctor placed her on strict bed rest. The clinic fired her. This would be illegal had the clinic been a federal operation, but it was an Arizona-based concession, and it is legal in Arizona to fire women who become pregnant and are unable to work.

We had to leave Grand Canyon Village because the only

housing available was for employees of the clinic. With Connie on strict bed rest, I moved us to a temporary place from which we could find a real house to buy. Within a month we found a fantastic place, but we needed a $40,000 down payment. We did not quite have the money. To move out of the dinky, dingy little hole we were living in, the one with the shared bathroom down the hall, we needed that money, so I had to return to the river to work this trip.

It had not been an easy decision, but here I was. Our contingency plan, should Connie go into labor, was to phone the Park Service, advise them of the life-threatening aspect of her pregnancy, and then evacuate me to the Flagstaff Medical Center. All of this could happen in a matter of a few hours, provided that a relief boatman could fly in to row my boat and passengers downriver. While not much of a plan, we were stuck with it.

I scanned the sky all the time, but now that we were an hour from Phantom Ranch (Mile 88) it looked as if I would get some concrete news over the phone.

The long tables in the Phantom Ranch canteen were studded with the elbows of red-faced, sweating tourists gulping expensive lemonades and beers. The eyes of several reminded me of trapped animals. I could read their thoughts, "How am I ever going to climb back up that trail?" It was 105° F in the shade that afternoon, not a particularly favorable time of day to hike. A few of them were unwrapping chocolate bars—carefully, as if the chocolate had been handed out as sacred food by some guru. Behind me, the air conditioner above the door whirred, fighting a last-ditch stand against the shock waves of a solar supernova. It was winning. The temperature in here was thirty degrees cooler than what existed outside the door. Only a half mile from the river and locked between walls that absorbed heat like a sponge, Phantom Ranch seemed to me a civilized slap in the face. But at the moment this slap was okay: they had a touch-tone phone outside near the public rest rooms. I checked the boatmen's mailbox and then made a beeline for Ma Bell. As I walked beneath the valiant air conditioner to exit the canteen, a blast of unnaturally cool air boiled behind me into the atmosphere.

"So, how are you doing?" I asked when Connie picked up the phone. The standard joke between us was that she would deliver the world's largest baby because it refused to be born.

"Okay, but you better get up here," Connie advised with unusual urgency in her voice. "Doctor Hampton is going to C-section me tomorrow morning at seven thirty."

"Okay. Does Walker have someone coming in to replace me?"

"Stanley decided to do us a favor. He's passing on that Dolores trip he and Kris were going on. There's nobody else available. Walker tried all over."

"Great. I owe him one. How's he getting to the rim?"

"Kris will drive him. Then she'll wait for you."

"Then I better get my pack and get moving."

"Yes. I want you here."

"I'm on my way."

"How hot is it down there?"

"One hundred and five in the shade." (But there was no shade.)

"Drink a lot of water."

"I will. See you tonight."

I threw my essential stuff into my large day pack and tidied up my boat for Stanley. I said good-bye to Mike and my passengers and then to everyone on the trip, made sure Boyle would keep Mike working with the crew, and then promised to return by hiking down Havasu five days later on day ten (we still needed the money, and Connie would have her best friend, Marian Spotsworth, who was also a nurse and who lived down the hall, to take care of her). Karen, Chris, and Melanie doubled up on the other boats because Stanley would not reach the "Boat

147

Beach" here at Phantom until after nightfall, and when he did he would sleep on "our" boat and then, at dawn, row downstream two miles to camp to join the trip.

After going over all the details a final time with Boyle, the trip leader, I shouldered my pack and started the ten-and-a-half-mile hike up to Grand Canyon Village. The trail was hot. Real hot. It did cool off after nearly three hours—when I had climbed about 4,000 of the 4,600 vertical feet. This was where I met Stan Boor walking down the trail in a mile-eating gait and listening to his Walkman. He was wearing a Sobek Alas Exploratory T-shirt. Stan and I had run a paddleboat together on the first descent of the Alas River, which roars out of the wilds of northern Sumatra. The Alas had been the most difficult river either of us had run. Now we were friends. I thanked him again for the favor he was doing me and wished him great runs in the challenging section of river below Phantom. He said "no problem," his positive attitude rising, as usual, to the fore. Twenty minutes later I reached the rim and rendezvoused with Kris.

At exactly 7:30 the next morning, Hampton sliced a deft bikini cut across Connie's abdomen and delivered an 8 pound, 2 ounce boy. I had lost half that weight myself through dehydration fourteen hours earlier and was gulping water now at every opportunity. Connie came out fine, considering. Little what-should-we-name-him was in perfect shape.

What *should* we name him? He was most likely conceived on the South Rim, or possibly down here on the river. We agreed that his name should reflect his Canyon origin, but the choice of Canyon names was terrible. We could not name him Coconino or Kaibab or any of the other local geological names. They were too weird and the sixties had been twenty years ago. What do you name a baby boy conceived on the most famous cliff in the world?

We named him Cliff. It was easy. But Connie figured he needed more. Because he had been so late and so mysteriously quiet, we had been calling him Cliff Hanger for a month. But no

148

child should be burdened with that. All his life when he wrote Cliff H. Ghiglieri people would ask, "What's the *H* stand for?" and he would have to explain it. But Connie liked that *H*. We groped for a significant and euphonious regional name beginning with *H*.

Old John T. Hance, the first historical resident of the South Rim, suddenly became far more than just an historical figure. He became a namesake for Cliff Hance Ghiglieri.

The birth of Cliff Hance Ghiglieri involved more pain than we knew. Later that morning Stan flipped "our" boat in Crystal Rapid, sending himself and four passengers shooting downriver on a swim for their lives. At the foot of the rapid Fabry, Boyle, and Craig plucked them out.

10

The Aisle of the Conquistadores

Heat waves shimmered, rising from the Tonto Platform a hundred feet above the river. The long, straight corridor of ledgy Tapeats Sandstone ahead was topped by orderly cliffs and terraces of the Paleozoic Sequence extending upward both north and south for distances that seemed too gigantic to be real. They seemed a trick of the camera. But there was no camera, just our eyes. Here the river has sliced a mile deep into these red and tan rocks to create an immense, arrow-straight corridor within Grand Canyon. And even after rowing 120 miles of the Colorado, when I first gaze down the Aisle of the Conquistadores it still awakes the primitive in me, like a three-alarm fire.

We had just hiked back into the scorching sunlight from the cool, dark, vertical depths of Blacktail Canyon, marking the head of the aisle. Besides offering relief from the heat, Blacktail's footing is the ultimate bedrock: polished, gnarled Vishnu Schist and Zoroaster Granite contrasting with the orderly, friendly Tapeats at eye level. It is separated by the 1.4 billion years of the Great Unconformity as if it were planned that way by some illustrator of textbooks. Already I regretted leaving its vertical waterfall dripping into a narrow emerald pool walled with lush maidenhair ferns. It was hot out here.

Squat barrel cacti stood on the Tonto Platform, scattered at intervals like petrified sentinels witnessing a never-ending series of summers. To these aging plants, many of which were more than a century old, our five bright yellow boats floating down the Aisle of the Conquistadores was less than the flight of a gnat. Sprinkled among the patient barrel cacti were Utah Agave, some in their death blooms, plus banana yucca. Although this corridor was a sun-baked desert, it held life, and not merely the tenacious life of the Sonoran Desert zone but life in abundance.

Yet I have never understood why *this* corridor was named the Aisle of the Conquistadores. The name is strictly romantic. True, the conquistadores did "discover" Grand Canyon—after being led to it by Hopi guides. But no conquistador ever gazed into the depths of the Canyon here in this stretch named for them. No conquistador came within fifty miles of here.

To me this was the Aisle of the Desert Bighorn—the single greatest stronghold in Grand Canyon of these vanishing symbols of the Southwest. And, as usual, as I rowed downstream here I scanned the Tonto Platform atop the Tapeats Sandstone. But the only things moving up there were heat waves shimmering off the rock.

Around March here, most ewes drop their lambs. Through spring and summer the ewes travel with a few related females in search of the most succulent forage—which, during the dry season before the monsoons, is found up side canyons near springs and seeps. And because the ewes will not mate again until September or later, the rams spend this time wandering with one another, or alone, higher in the tributary canyons. By August this changes. The monsoons water grasses here in the inner corridor and attract sheep. Mature rams gather and defend harems of ewes, while lesser rams bide their time until they can challenge and usurp a harem whose ram is too exhausted to put up a good defense.

As I rowed, I studied the slopes and terraces atop the Tapeats as if they sheltered the last desert bighorn on the planet—which is not as far-fetched as it sounds. When Meriwether Lewis and William Clark were trekking across North America and back, between 1.5 and 2 million bighorn sheep of all species roamed the West from Mexico to Alaska. Today only 15,000 to 20,000 desert bighorn survive (roughly two thirds of them in the southwestern states, and the rest in northern Mexico). While those surviving here in the Canyon are not the last on earth, I suspect they may be the most important.

Why? Even now, despite many expensive transplant projects, desert bighorn are vanishing from the Southwest. Of those 15,000 to 20,000 survivors, about 2,500 are in Arizona, and possibly 500 of them are here in the Canyon. Their decline is due to the invasions of Spaniards and Anglos, who waged war on them—both conventional and biological: through overhunting, competition with feral burros and livestock, range destruction, and worst of all, poisoning their ranges with numerous stock diseases spread by grazing livestock and to which bighorn have no immunity. Today most national forests and Bureau of Land Management lands in the continental United States are leased to ranchers to graze livestock. Especially in the past, domestic sheep set loose on these lands carried European diseases lethal to native bighorn. They died like flies. In some regions today lamb mortality from these imported diseases is still close to 90 percent. The competition that bighorns weakened by disease face against the cattle, sheep, horses, and burros that nibble everything green in their habitat down to the roots sometimes becomes the coup de grace that eradicates bighorns completely. In regions where they do somehow survive both competition and disease, overhunting often wipes them out.

These bighorns are unusual because domestic sheep never grazed here in the heart of Grand Canyon; hence they may be a special population free from exotic diseases. Further, because no bighorn have been added from other populations, this may also be the most genetically pure population left in the world. These possibilities kindled in me, years ago, the goal of discovering the population dynamics and other ecological relationships of the desert bighorn here. The park had granted me permission to start, but money had proved more illusive than the rare bighorns themselves.

Timing was important. Around 1900, prospector and tour guide William Wallace Bass had brought two railroad cars of burros and released them into the Canyon at about Mile 108 as breeding stock. Prior to 1980, these feral burros multiplying in the Canyon ate half the food favored by bighorns and, because of their size and their possession of upper incisors capable of "nibbling" anything that grew here, could out-compete them. Because the National Park Mandate prohibits allowing alien species to invade and multiply—especially at the expense of native species—the park had been shooting feral burros—2,600 of them between 1924 and 1978. But in the late 1970s a stink was raised over murdering these innocent little friends of the prospectors. The park could not afford to round up each burro, but a private group, Friends of the Burros, raised $600,000 to pay for airlifting 600 of them out by helicopter. By 1981 all but an elusive one or two were gone, and the desert bighorns were free of competition. I had wanted to see the repercussions of this.

We drifted down the corridor between distant, immense cliffs. Yes, I'd seen bighorn there on that slope—and there on those ledges too. A couple of albino ewes, one old and one young but in separate matriarchies, ranged here on the south side in late summer. I had seen them several times, usually in some ram's harem, and had approached to within twenty-five yards of them. Since the 1919 ban on hunting in Grand Canyon National Park, the sheep have become blasé about people as long as we remain lower on the terrain than they are. A person approaching from above will trigger a dead run. Instinct has imbued bighorns with a must-be-above psychology because their main means of escape from predators is ascending rough terrain or a broken cliff in a hurry. Alaskan bighorn (Dall sheep) calmly watch wolves below them because escape upward is certain. But a wolf on their level or above spurs instant panic.

Be that as it may, I saw no bighorn here. The Canyon walls seemed to close in a little, but my eyes refused to stop searching the low ledges of Tapeats as we glided down the Aisle of the Conquistadores.

Forster Rapid. Still no bighorns. I set up on the tongue and then, during the entry, scrutinized the mouth of Forster Canyon. I had spotted bighorns here before—early enough to catch the eddy and get photos and good data on harems: membership by age and sex.

Now, at the last second, I pivoted the boat to slap into the first peaking wave. As we dropped into the trough I scanned the slope again, grudgingly parceling out the minimum time for rowing. My three passengers in front did not notice that I was driving almost blind. But still no bananas. This made me look even harder. We exited the Aisle with a ninety-degree bend to the north and soon arrived at Fossil Rapid—still a hot spot for desert bighorn. I let the river accelerate us to where I wanted us to go, then I devoured the slopes with my eyes.

Water pounded over us like a slapstick joke. My fault. I pivoted to straighten us and finally admitted to myself that I had to give more attention to piloting my boat. Despite this, my eyes strayed back to the left slope.

Four ewes plus two tiny lambs and two yearlings (born last year) were grazing on sparse desert grass scattered on a huge sand dune. Pay dirt. The eight survivors of the Pleistocene gazed at us placidly as our other boats joined us in this eddy below them. But their placidity was misleading; ewes with lambs but without a ram nearby are skittish about people on foot.

"If you want a picture, you might be able to get a little closer," I advised. "But take it slowly."

Across the boats, surplus Vietnam-era ammo cans popped open as if the word had just come in that our perimeter had been penetrated and we were about to be overrun. Instead of shiny brass rounds destined for M-16 magazines, cameras emerged from these cans, were aimed at the timid sheep, and clicked repeatedly, zapping them with invisible rays, stealing their souls but making them future celebrities in living-room slide shows.

T. A. jumped to the sand and advanced for a close-up. When he had closed to half the distance, the ewes startled, spun, and climbed higher. Now with each step he took they retreated at least as far. Soon they vanished beyond the crest of the dune and ended our interview. I jotted my notes in my river log for that scientific report I was planning, then repacked my binoculars and shoved off to row us into the second half of Fossil Rapid.

As bighorn sheep become more rare, they become more valuable. But what's strange is that they are not as much more valuable to save as they are more valuable to *kill*. Bighorns are so rare that hunts, legal or not, are big business. The Hualapai Tribe auctioned ten permits in 1986 to hunt a ram on their reservation in western Grand Canyon. One sold for $35,000, and that's not even close to the record. A man in Montana recently paid $93,000 to hunt a mountain bighorn ram.

The strange—and virtually unknown—changes in the lives of the desert bighorn sheep in Grand Canyon led me to reflect on how wildlife management in and around the Canyon had gone through its own dark age as backward as anything in medieval Europe. In 1903, Teddy Roosevelt saw the Canyon for the first time and said it was "the most impressive piece of scenery I have ever looked at," adding a warning for posterity: "Leave it as it is. You cannot improve on it." In 1906, Roosevelt signed the bill that designated it as Grand Canyon Game Reserve. But in direct contradiction to what we would expect, the act launched an American pogrom on the wildlife here.

In those days the term *game animals* did not include predators. On the contrary, predators—*all* natural predators—were "enemies" of game animals. Roosevelt's signature provided federal funding to hire full-time predator-control men armed with traps, poison, rifles, shotguns, hounds, et cetera to protect the estimated 4,000 deer on the Kaibab Plateau. Between 1906 and

1931 these government hunters trapped or shot 781 mountain lions, 30 wolves (all that existed), 4,889 coyotes, 554 bobcats, and hundreds of birds of prey. With their natural controls gone, the deer herd exploded. Then crashed.

By 1924 the herd had grown, by one estimate, to approximately 100,000—a 25-fold increase. The problem now was that the habitat, in its original lush condition, could sustain only about 20,000 deer. During the following winter, starvation and disease wiped out nearly 50,000 of them. Even so, many does bore fawns, and the population climbed to roughly 70,000 deer in spring. During the winter of 1925–26, half of them starved to death again. Corpses littered the Kaibab Forest. Ravens pecked at many of them, but most of the natural scavengers were gone, so many other corpses simply rotted when spring came. And the forest itself was decimated. What had previously been the most lush habitat in Arizona had been destroyed by ravenous deer eating nearly every bit of plant material from the forest floor to as high as they could reach while standing on their hind legs. By 1931 about 20,000 deer remained in a devastated habitat. By 1939 there were only 10,000 deer. Oddly, Arizona Game and Fish retained an open season on predators here even after the federal hunters were pulled out.

Game management on the North Rim of the Canyon after the turn of the century could not have been much more primitive—unless they handed out machine guns free of charge. A few years earlier, exploration of the river itself was even more so. Whenever I float this relatively benign stretch of the Colorado through the Aisle of the Conquistadores—flat water allows a boatman the free moments to contemplate literally anything—I think of the second expedition to run the entire river through Grand Canyon. Maybe this is because by the time they too reached near the halfway point in the Canyon, they were equally glad to be alive. The survivors certainly had reason to be.

The Brown-Stanton Expedition was not just a major event in the exploration of Grand Canyon, it is one of the clearest stories on record of the Great American Dream turned to night-mare. This exploration was attempted in 1889 and was the first attempt to follow Powell's expeditions twenty years earlier.

But this expedition began with a dream even bolder than Powell's ambition of navigating the river and filling in blank spaces on the map. In 1889 a real estate investor among Denver's upper crust named Frank Mason Brown was seduced with an idea presented to him by a prospector named S. S. Harper to build a railroad within the inner gorges of the Colorado. Huge deposits of coal in the Four Corners region could be sold in San Diego, and manufactured goods from the West Coast could be sold profitably in the growing Southwest. A railroad within the canyons would avoid mountains. The idea hooked Brown, who imagined it to be a catapult to fantastic wealth. He magnetized additional investors and became president of his infant Denver, Colorado, Canyon and Pacific Railroad Company.

Brown hired Robert Brewster Stanton, an accomplished civil engineer, as his surveyor. With a dozen hired hands, they set off from Green River, Utah, in five boats Brown had designed. He had interviewed Powell but had disregarded the Major's warning that strong boats of oak plus life jackets were vital. Instead, Brown decided on light boats of brittle red cedar only fifteen feet long, forty inches wide, and eighteen inches deep—glorified canoes lacking even a superficial resemblance to Powell's boats, which themselves had barely survived.

Stanton, expecting twenty-one-foot boats of oak, became alarmed when he saw Brown's boats, two of which had split just during shipping. Worse, when Stanton asked him to hire boat-men, Brown said two lawyer friends and he would row three boats, the men the others. Worse yet, when he begged Brown to buy life jackets, Brown said they were unneeded because the men would walk around the rapids. Stanton considered Brown an op-timist. Both were dead wrong.

Only a fraction of the expedition's supplies fitted into Brown's little boats, so they built a raft of containers to hold the rest. They towed it more than a hundred miles down flat water, then lost it above Cataract Canyon. This loss augured hard times. For seventeen days the surveyors bungled their way down less than forty miles of wild river, demolishing two boats, damaging the others, and losing most of their food, much of their personal gear, and a lot of surveying equipment, and ending with a mutiny by the starving men. Brown sidestepped this by rowing the ten mutineers in the three surviving boats to Hite, abandoning Stanton and his four loyal assistants to survey for three more days. Brown hired men to row back upstream and pick them up.

This may sound like roughing it, but compared to what happened next it was a lark. By the time they reached Lee's Ferry, Brown's crew had dwindled to eight. Brown's lawyer boatmen absconded to Denver. The men reprovisioned, then entered Marble Canyon in the three patched-up boats on a flow of roughly 30,000 cfs. They portaged Badger and Soap Creek rapids.

In the morning they shoved back onto the river. With Brown rowing, Harry McDonald and he caught the current first, and because the next boat containing Stanton, Peter Hansbrough, and Henry Richards had gotten hung up on a rock near shore, Brown's and McDonald's lead boat gained a lead of minutes. This distant spacing between the first two boats would have lethal consequences. Many of us have learned this lesson the hard way, although not quite as hard as they did.

A quarter mile below Soap Creek, the river slices into the Esplanade Member of the Supai Formation, a sinuously carved sandstone through which the river narrows and gains speed. At the mouth of Salt Water Wash is a riffle walled by ledges of Esplanade. At 30,000 cfs it is almost a rapid. Brown and McDonald crossed the powerful left eddy fence. Their little boat flipped instantly. McDonald was tossed into the downstream current, where he swam for his life and managed to haul himself out of the river a few hundred yards downstream.

Brown was dumped into the eddy fence, where whirlpools between the main current and the eddy cycle endlessly. He could not swim out. He probably made the mistake of yielding to his natural inclination to try to swim the fifty yards to shore against the strong eddy current going upstream. Instead, he should have stroked the five yards to the main current and then downstream to follow McDonald.

Meanwhile, McDonald sprinted up the left shore on ledges of Supai, stopping to crawl around abrupt cracks yawning too wide to leap, and shouted over the roar to Stanton's boat that Brown was caught in the whirlpools in their path. But when Stanton's boat closed to within 200 feet of Brown, who was now struggling weakly, McDonald saw him vanish under the silty water.

As Hansbrough rowed to the spot, Brown's notebook shot to the surface. Stanton grabbed it. No one ever saw Frank M. Brown alive again. If only he had spent those few dollars for life jackets.

Brown's death did not end the story; it only concluded a chapter on poetic justice. The seven men searched the eddy and the riverbanks for Brown for the rest of the day. Peter Hansbrough carved an inscription in inch-high letters still visible to-day in the black desert varnish in the angle of the Esplanade directly across the eddy from where Brown vanished—a few yards from that elusive bit of shore that he had been swimming toward so helplessly: "F. M. Brown, Pres. D. C. C. & P. R., was drowned July 10, 1889 opposite this point."

In his later account, Stanton mentions no thought of abandoning the expedition. Instead, with a bulldog's persistence, the seven survivors continued downstream. But their flimsy boats, lack of experience, and repeated lining and portaging cut their progress to only three or four miles a day—about what we row in an hour.

Five days after Brown died, the men had lined their boats along the edge of a rapid, most likely Twenty-five Mile Rapid. One boat had run the lower section and was waiting below as a safety net.

Hansbrough and Richards climbed into the second boat and

rowed into the tail waves of the rapid. The waves repulsed them back toward the left into an overhanging cliff in the Redwall Limestone, where their boat flipped. As Stanton screamed to the men waiting in the boat below them, Hansbrough and Richards struggled in water thick with silt due to recent monsoonal storms. The silt so loaded their clothes that they were dragged under despite their efforts. They sank like stones even before reaching the rescue boat pulling out from the beach a few hundred feet farther down.

Again life jackets would have saved these men. This was the last straw for Stanton and the others; they decided to escape the Canyon at their first opportunity. For the next two days the five survivors ran scared, lining or portaging everything remotely resembling a rapid, until they reached South Canyon at Mile 31.5. They consumed a week traveling less than the section we normally row on day two. They spent the night at South Canyon crouched in a limestone cave and sitting through flashes of lightning and cracks of thunder that reverberated off the cliffs during one of the wildest monsoonal storms Stanton had ever experienced. The exploding sky must have seemed a message from the gods themselves that the river was not to be trifled with.

In the morning they found some Anasazi ruins and a route out to the west. They decided to try it. While caching their goods in that cave well above the high-water line, Stanton spotted a bundle floating down the river. It was wearing Brown's coat. Two men ran to a boat and Stanton ran along shore, but the body had too great a lead and vanished down the Colorado. They turned back to hike out.

Although a normal man would thank his lucky stars for having escaped an experience like this, Stanton took the whole package as a personal challenge.

The five climbed out of South Canyon, then hoofed it across the Kaibab Plateau. Every dollar of the expedition had floated into oblivion on Brown. Stanton borrowed money for the railroad trip to Denver, where he inveigled a group of local investors into financing a new expedition. Even S. S. Harper, the idea man, bought in again. Stanton added $12,500 of his own money, then ordered three twenty-two-foot oak boats built to his design with multiple airtight compartments, plus special life jackets, also of his own design. In less than six months he was back on the water—this time on only 5,000 to 6,000 cfs.

The most amazing part of his comeback was that three of the other four survivors returned with him as members of the second crew. One, the photographer Nims, took a nasty fall upstream of Mile 20 and was carried and dragged unconscious by rope up the cliffs of Ryder Canyon to the rim in freezing weather. Stanton refused to pay Nims or recompense him for his travel or medical expenses because the photographer had been taking a "scenic" shot of the men when he fell—not a strict survey shot as Stanton had instructed.

At Mile 44, Stanton's crew found Peter Hansbrough's body and buried it. They named the prominence looking down on his grave Point Hansbrough.

Ninety-nine miles down the Canyon a second survivor, McDonald (Brown's boatman), became disgusted with the work Stanton expected for paying him double. The straw that broke McDonald's back was cruel. Near Mile 81, Stanton's crew struggled to line their boats down the edge of Grapevine Rapid (although Stanton thought he was at Sockdolager Rapid at Mile 79. The *Sweet Marie* broached against a boulder, sank, then postage-stamped to it. When Herculean efforts failed to retrieve it, they resorted to securing it with ropes and hoping that the flow would change. Overnight the river rose. The next morning when they hauled the *Marie* out they discovered a hole in it big enough to crawl through.

McDonald figured he could repair it. For five days the expedition camped, wet and freezing, on that narrow boulder bar where the sun rarely hits in winter, while McDonald and two helpers cut four feet out of the damaged middle of the *Marie* and carefully mated her ends together. Scraps from the boat provided their firewood for cooking. Once finished, the *Marie* floated like new.

They continued downriver and lined numerous rapids that they could have rowed easily if they had had the skill. Days later Stanton's crew stopped with their hearts in their throats at Horn Creek Rapid (Mile 90). They lined the first boat down the boulder-strewn cascade, but only sheer luck prevented their losing it.

Stanton concluded that they should try something different with the second boat—"ghost boating" it without lines. They shoved the *Marie* into the current. She dropped into the rapid and zigzagged through part of it but struck another boulder and sank. Then, as McDonald and the others watched, the *Marie* resurfaced in pieces that Stanton claimed were no larger than toothpicks.

McDonald then told Stanton the job was not worth even double pay. Eight miles downstream (two dozen miles upstream from where I was now in the aisle of the desert bighorn) he said good-bye and hiked out alone from Crystal Creek to the Kaibab Plateau, a very tough climb in winter. Then he plodded for days through record snow to a lineman's shack and then to Kanab, Utah.

The rest pressed on. Seven of Stanton's twelve starters and two of his boats made it all the way through Grand Canyon, becoming the second trip ever to run the entire Canyon—this time with no deaths. Due to the careful surveying and dozens of portages, it took them eighty days. Stanton completed his survey and concluded that a railroad should be built, but he never persuaded any railroad magnates to back the scheme. And despite Stanton's obsession, the Denver, Colorado, Canyon and Pacific Railroad never laid track. It's a good thing for the bighorn sheep and for us that it didn't.

11
Thunder

I n Grand Canyon, where even superlatives fail to match the real scale of
things, it's hard to explain in mere
words how and why a special place is
special. Normal earthly experience does not prepare people to
grasp the scope of things here without seeing them—and then
attempting to climb them. Knowing this makes it even tougher.
Even so, it is worth a try.

I have a favorite place in Grand Canyon. It's not a secret
place, although sometimes I wish it were. Anasazi Indians knew
of it a thousand years ago. Hunters of the Desert Archaic probably were familiar with the region millennia earlier. People of
both groups could descend into the Canyon and hike to it along
a route from what we today call Monument Point on the North
Rim, a dozen miles from (and a mile above) the river. Paiutes
followed the Puebloans several centuries later. Prospectors used
the same route just a century ago but found no mineral wealth.
The U.S. Forest Service improved parts of the trail in 1926. So,
rather than being secret, my favorite place is more like Grand
Central Station in slow motion. But it holds the most incredible
collection of landscapes in Grand Canyon.

The route to it from the river begins at Tapeats Rapid (Mile

134) and circles above the roaring gulch of lower Tapeats Creek along a steep trail ascending a few hundred feet straight up to a bench atop the Diabase Sill, splitting the Bass Limestone. This part of the trail is so steep that climbing it becomes an act of faith in the work ethic. Nothing this taxing could fail to earn a reward. And it does. The view is stunning. From up here our boats look like yellow toys dinky enough to drop down the neck of a water bottle. And Tapeats Creek—at about 65 cfs the largest creek originating in Grand Canyon and loaded with trout, the kind of creek that can knock people down when they misstep crossing it—seems a cellophane-thin trickle where it joins the implacable beast of the Colorado. From up here atop the Diabase Sill, it is a mere saunter (a careful one) along the bench descending gently inland to Tapeats Creek.

What lies ahead is hidden by a bend in the canyon, but it is so huge that one can safely anticipate that something interesting awaits discovery up there. The question is, what? Here the hidden canyon becomes an urgent mystery, a promise, a mysterious desert oasis concealing austere delights to make the senses whirl. Soon the route connects with the creek where it metamorphoses from a roaring gorge into a trout stream again, tumbling over waterfalls flanked by cottonwoods. It races to the Colorado as if the earth were about to end and it had to be there on time. The best crossing is located at a wide ford where no rope is necessary unless the creek is in spate. From here the path snakes upstream between groves of succulent prickly pear cacti and junglelike thickets of bear grass hiding rattlesnakes, and over ledges perched above the rushing creek. Soon it enters a wide canyon flanked by tall cliffs of Shinumo Quartzite, cliffs of swirled maroon and purple and tan and cream.

In the Shinumo Quartzite are sheer cliffs punctuated by ledges, niches, overhangs, benches, caves . . . and dozens of hidden granaries of the Anasazi. One, the Mystic Eye, is unique in Grand Canyon. It is constructed entirely of wattle and daub (straight sticks completely plastered with mortar) except for its threshold and door, which was a solid slab of quartzite. It is also perched high in a cliff face, in a niche better suited to a nesting peregrine falcon than a granary. Because of an exposure of from thirty to more than a hundred feet of freefall, the difficult friction traverse across the cliff face required to visit the Mystic Eye stops normal humans cold—even some abnormal ones. Boatmen have fallen from this cliff while attempting to traverse the quartzite, paying the price of trying to absorb a bit of the magic of the Eye. Mystic Eye manna, at too high a price. But most Canyon boatmen do not even know where the Mystic Eye is, and it is far too dangerous to take clients there. So it remains somewhat of a secret. Another secret is how an Anasazi loaded with maize made this move across the cliff. Maybe they built ladders. The view from the eye is worth the risk—but only if you are very good on cliffs.

The path in the valley winds between these cliffs and passes several dwellings of the vanished Anasazi, but to the untrained eye most of these ruins are invisible. The Indians must have planted fields of maize, beans, and squash along the expanse of gentle slopes near the roaring creek. When I study the lay of the land here, I am forced to admit that agriculture would have been possible with more rain, but converting the boulder-strewn slopes to terraced fields would have been hellish work. Yet the Anasazi did just this.

Less than three miles from the Colorado, we abruptly abandon Tapeats Creek and the swirled cliffs to ascend a steep trail to the northwest—we are now traversing a slope parallel to the world's shortest river.

Thunder River is a spiritual experience. No one who crawls up the switchbacks on that sun-baked slope, then suddenly beholds an entire river blasting forth from the dry face of the Redwall Limestone fails to be moved. No matter that a half mile from this roaring spring and tumble of wild cascades Thunder loses its identity by merging with Tapeats Creek (is this the only case of a river flowing into a creek?), Thunder River is unforgettable, and because of the struggle to reach it, precious.

The temperature among the cottonwoods along the falls a hundred feet below the source is perfect. Thunder River pours over mossy boulders and the twisted and gnarled roots of cottonwoods. Crimson monkey flowers form a foot-deep carpet flanking the flow. They are in turn bordered by lush maidenhair ferns, horsetails, then willows, grasses, and herbs. The place is a riotous jungle oasis perched on the face of a bone-dry cliff in a barren, sun-baked canyon.

The water tastes pure and cold and seems a sacrament. A glance upward to the narrow slices in the limestone from which Thunder Spring blasts into the sunlight to become Thunder Falls and then Thunder River gives no hint, though, that through those narrow portals and within that massive cliff are underground falls and grottos, sluices and caves, rooms and auditoriums, rapids and lakes . . . and fear, in helpings not normally available in the outside world.

On my first hike up here in 1976, I climbed the cliff alone past several nasty exposures to gaze into the source of Thunder. I don't recommend it. The river blasts out of a four-foot-wide fissure with the speed of an F-16. Or so it seems. And to enter the cave one must step over this Mach-1 blast of water onto a steep, wet wall only a few feet from the falls. The thought alone prompts panic. This move into Thunder Cave makes the friction traverse to the Mystic Eye seem like rocking on a front porch with a mint julep. A misstep would join you with Thunder Falls and spin you into a mix of air and falling water for a couple of seconds before you pound onto the rocks below. From there you would rag-doll down one impressive and scenic set of falls to another. Altogether an unattractive prospect.

I gazed into the cave and felt fear prowl around in me like a loose tiger. I decided not to try that one step over Thunder River. Even so, I could not shake the conviction that the secrets of the universe were contained within that unattainable cavern. It bothered me for nine years, during most of which I led people on trip after trip up Tapeats Creek and up Thunder River to the falls below the source. The awe, joy, reverence, and appreciation people felt did not quite satisfy me. None of them imagined that a cave existed behind that cosmic fire hose in the cliff. But I knew, and it seemed to stare down at me in patient scorn: "I'm here."

Nor could I simply absent myself briefly from my group and hop up there with a flashlight and rope to explore the cave in the mere hour or so we spent there. First of all, an hour is too little time. Second, someone would want to come with me. Taking any paying client up there would be potential manslaughter. Yet, although I don't recommend it to anyone, that hole in the cliff haunted me.

Finally in 1985, on a trip I was leading, Michael Boyle, Mike Anderson (our trainee), and I left our group early to hustle up to Thunder Spring, where we climbed the cliff face, slipped on wool shirts and headlamps, uncoiled our rope, belayed one another on the entry, and made that horrendous step across the rush of Thunder River. "One small step for a man . . . " and one scary one. But with the rope for protection, it was easy. None of us slipped. We were inside Thunder Cave. Immediately at our backs an entire river shot into the bright sunlight and thin air. We deroped ourselves. A slip here now would mean adios.

The long entry into Thunder Cave is merely a deep slit with a river rushing along the bottom. For the next half hour, each of us employed every bouldering trick we knew to straddle the flow as we traversed our way along the vertical Redwall Limestone. Here the rock was eroded into knife blades at nearly every edge. There was literally nothing in this cave that was smooth or rounded. Carefully, cautiously, and with the precision of diamond cutters, we examined the cliff walls for tiny cracks and nicks that would allow a hand- or foothold. It was as if we had a hundred-foot exposure below us rather than ten feet to a small river. Actually, we had both; one led to the next. Inch by inch and foot by foot, we penetrated the zone of mystery. Eventually, after several traverses, climbs and descents, chimney moves and crawls, we found sections where we could drop into the bed of the creek itself and walk or wade upstream. This was heaven.

Why do batteries always fade when you need them the most? Despite fading lamps we pressed on through grottos and stream corridors. We paused at waterfalls that would have been scenic in the world behind us, then climbed up around them to continue. Would we discover something absolutely unbelievable beyond the next bend? A cavern large enough to support the last survivors of a nuclear holocaust? The skeleton of some antediluvian beast . . . a *Tyrannosaurus rex*? A portal though the fabric of time to the lost world of the Anasazi?

What we did find was grottos, some as big as auditoriums, alternating with narrow slits through which Thunder River raced. We found falls, walls, and dry "extinct" caves branching from this main channel and begging to be explored. We also found ourselves shivering.

Finally we came upon a silent lagoon, a wall-to-wall lake. Upstream the roof of the cave dipped almost to the water. It did not look inviting. We were cold. But it did exert an irresistible magnetism. One after another we slid into the water and swam upstream. We climbed out beyond the dip and, shivering, studied the lake. It continued for an unknown distance, hidden by the very low roof. We needed a small boat or an air mattress. Reluctant to leave but satisfied with our efforts, we turned and retraced our steps. Our lamps were glowing only a dim yellow now.

We rounded one of the last sharp corners. Here we had to straddle the stream channel about six feet above it with hands and feet on each side. Boyle was just ahead. Anderson had fallen behind. I looked back and could not even see the glow from his lamp. I started worrying about him. I tried to find a more comfortable position in which to wait, but half-inch-wide nicks in the Redwall were all that was available. I finally spotted his lamp, then him. Then I turned to follow Boyle. As my weight shifted, the tiny ledge of rock under the heel of my right hand broke off, and I suddenly lost that hold plus my other two contacts. Abruptly I free-fell into Thunder River in the dark. My knee jammed into the sharp point of a large boulder. It hurt. I climbed back out. In the dim glow of my lamp, blood flowed down my shin.

The sunlight beyond the opening of Thunder Spring seemed like the flash of an atomic explosion. Each of us had to avert his head for a moment for our pupils to adjust to the brightness. Three hours after our entry, we roped up on belay again and made that single step across Thunder to Grand Canyon. It was easier this time.

Everyone had gone. We descended the cliffs to the trail. I looked again at my knee. It was as swollen as a ripe peach. Then I looked back up at that enigmatic slit in the Redwall with the explosion of water bursting from it.

Thunder Spring is an amazing reality. It seems too improbable to invent in fiction, and it is not clear to everyone, even when seeing the roaring fountain in the face of the cliff, exactly *what* Thunder Spring is. Fellow guide T. A. told me incredulously about one woman who never did grasp it.

"How does this water get here, T. A.?" she had asked.

"Well, there's an aquifer in the Redwall Limestone that's impermeable and collects groundwater moving into it from the Kaibab Plateau. The outlet for that aquifer is right there—Thunder Spring."

"Where does all that water come from?" she pressed.

Not thinking that her question was serious, T. A. answered, "Every two thousand years a spaceship comes here and fills it up."

"No, really, who fills it up?"

Finally realizing that the gap in understanding was too great, T. A. told her what she wanted to hear, "The Park Service."

"I thought so," she said smugly.

My knee was sore, the price for information. It was worth it. At least most of the four miles ahead of me were downhill.

But a trail continues from here uphill. About five hundred vertical feet higher, this trail tops out to traverse a broad slump called Surprise Valley. Millions of years ago a cubic mile of rock, the largest known landslide in Grand Canyon, collapsed from the cliff face and slid south to dam the Colorado and divert it. The top of this slump to the east, 2,000 feet higher than the river, is Surprise Valley. It offers an excellent hiking route between here and the next huge tributary canyon to the west, Deer Creek.

Halfway across Surprise Valley, the trail forks to divert upward and northward along that old Anasazi route out of the Canyon. Once, on a lark, I led four hardy hikers up that trail a couple of miles to the Esplanade member of the Supai (nearly 1,000 feet higher than Surprise Valley). The temperature was about 4,000 degrees in the shade. It was hot. Very hot. And not a lick of shade.

We swallowed our last drops of water on the way up. The only other water available was back at Thunder Spring or farther away at Deer Creek Spring. But the view from the Esplanade was fantastic. Near the edge of the cliff I discovered a Stone Age hunting camp with arrowheads and broken stone knives. The Esplanade begged to be explored, but by then dehydration was making our mouths feel like old boot leather. The blistering heat drove us down into the drainage of Deer Creek.

But before it did, the panoramic view from the Esplanade rubbed salt into the intellectual wound that we really do not know yet exactly how Grand Canyon was formed. Given everything else we know about geology, it may come as a surprise that no one is certain how it came to be eroded and carved as it is.

Sure, we know the basics: the Colorado Plateau uplifted, and the river or rivers sliced deeply into it as if they were watery laser beams. But beyond this the details become increasingly fuzzy.

Obviously Grand Canyon was carved by erosion, river erosion primarily, and by tributary erosion, slab failure, frost heaving, spalling, rock avalanche, talus sliding, gravity faulting, landslide, debris flows, river anticlines, cave solution—and all of these processes were set into motion ultimately by the river. But erosion is the only agreed upon part of an answer that remains both incomplete and a source of continuing debate.

John Wesley Powell, the first scientific explorer here, supposed, partly correctly, that the Colorado River in this region predated the uplift of the Colorado Plateau. When the land slowly rose, he explained, the river simply entrenched itself ever deeper in its ancient bed, like a layer cake shoved upward into a knife. When one looks at the Canyon from either rim, this explanation feels satisfying. But a closer look evaporates this satisfaction like moisture at noon and makes Powell's interpretation impossible.

Most geologists agree that the Colorado River probably flowed out of what is now Colorado—along either its present course or what has now become a tributary, the San Juan River, for instance—across the Colorado Plateau prior to 20 million years ago. Exactly how much earlier remains unknown, but it was at a time when this plateau was not much of a plateau. In fact, the landscape here was lower in elevation than the landscape farther south in Arizona. The reversal to the present relative elevations here began 17 million years ago and continued until about 5 million years ago. Most geologists also agree that the old river flowed into the region of Marble Canyon, carving its way through the softer strata atop the resistant Kaibab, then deeper through it and the Paleozoic sequence to cut Marble Canyon. Disagreement, however, erupts over where the river flowed after it entered Marble Canyon.

Surely it must have flowed out the other end, one might point out. But, no, it couldn't have. The problem is an extensive though infant rock a thousand feet thick at the western boundary of Grand Canyon at the Grand Wash Cliffs. Called the Muddy Creek Formation, it is a mere 10.6 million years old. Because it straddles the present course of the Colorado exiting Grand Canyon and because it contains no Colorado River gravels but instead was deposited at right angles to the later course of the Colorado, the problem it poses is perplexing.

The Muddy Creek Formation never could have formed during its tens of millennia of deposition directly on top of this major river. The river would have carried it away. So, obviously the river had to have cut through this formation only after it was formed. To complicate matters further, a formation barely half as old, the Hualapai Limestone, sits atop the Muddy Creek Formation. The Hualapai Limestone appears to be a deposition from a deep lake. Both formations shout that the river could not have been flowing through there prior to six million years ago. Therefore the Colorado River must have exited Grand Canyon somewhere else for approximately 20 million years. The geological mystery is to find this missing exit.

Few exits from Grand Canyon exist through which the Colorado might have escaped. Geologists have hypothesized that the river once flowed "up" the 20-million-year-old drainage of Diamond Creek and Peach Springs Canyon (Mile 226), then, millions of years later, through huge caves in the Redwall (which no one has ever found). Alternatively, the river may have flowed into Marble Canyon, but instead of entering Grand Canyon, it absconded at Mile 61.5 by flowing "up" the ancestral bed of the Little Colorado to the southeast to join the Rio Grande system and exit into the Gulf of Mexico. Neither hypothesis holds water.

Luckily, because so few potential exits from Grand Canyon exist, we are left with only one more solid option. This is Kanab Creek at Mile 143.5. It is no surprise that the third major hypothesis for where the ancestral river flowed out of the Canyon, which was proposed by Ivo Lucchitta, places its exit here. To understand why, bear in mind that during those years the Gulf of

California did not exist; the region south of Lake Mead was then higher in elevation than the Kaibab Plateau. The path of least resistance for a river as large as the Colorado would have been northwest into Utah and Nevada, not southwest to a Gulf of California that did not exist and through mountains that did. Kanab is one of the rare antecedent streams in Grand Canyon, meaning that it was here before, or at the very beginning of, the Canyon's first cutting. Also in this rare category are the Little Colorado, Cataract Creek (which is the primary source of Havasu Creek, which reaches the river at Mile 157), and the dry branch of Diamond Creek leading down from Peach Springs. But only Kanab leads northwest from Grand Canyon.

Another clue to the puzzle is the group of gravel beds north of the present Colorado on the Shivwits Plateau. Their most likely source would have been near Prescott, Arizona, far to the south of Grand Canyon. These reveal that at least one major stream flowed from south to north across this region, perhaps joining the ancestral Colorado as it flowed north, or, more likely, cutting an antecedent stream that the ancestral Colorado later captured and superimposed. The ancestral Little Colorado and/ or Cataract Creek (Havasu Creek), both of which apparently predate the Colorado here, may have been the source of these gravels. The Kanab carried them north.

Obviously, if all this is correct, all three streams comprised one river system with two tributaries that crossed the Kaibab Plateau from southeast to northwest before the Colorado arrived on the scene. In other words, the Little Colorado was probably a much larger river draining a mountainous region to the southeast that was much higher than the Kaibab Plateau of those days. It probably carved the first eighty-two miles of Grand Canyon from Marble Canyon down to the present confluence of Kanab Creek with the Colorado at Mile 143.5, thus excavating a "starter" channel for the later Colorado.

Is Lucchitta's hypothesis correct? Probably, but the candidate regions for the continuation of the northward Colorado in

Utah and Nevada have been compressed and jumbled by basin-range faulting. As if to conceal their geological secrets still further, many regions where the old Colorado may have flowed beyond Grand Canyon have also been buried since then under flows of lava. And, while Lucchitta's hypothesis seems the most probable solution to the mystery and no hard evidence controverts it, the story is not over. It only accounts for the eastern half of Grand Canyon.

Somehow, about 5.3 million years ago, western Grand Canyon was carved, including the dissection of the Muddy Creek Formation and Hualapai Limestone at its mouth, to cause the abandonment of that elusive old exit toward the north.

Roughly 5.5 million years ago, as land subsided along the notorious San Andreas Fault to open the Gulf of California, a stream must have formed, draining the western or southwestern edge of the Kaibab. During the past 5.5 million years, the western plateau uplifted another 3,000 feet. All three hypotheses mentioned above end the same way, with the old Colorado being captured by headward erosion of this new drainage toward the heart of Grand Canyon. Stream capture is a common geomorphological process, especially where vast but adjoining regions change paradoxically in elevation with respect to one another across a fault. So this final marriage between the eastern and western Grand Canyon could have happened as proposed by all three hypotheses.

The Colorado River escaped west to carve the western Grand Canyon abruptly. Six-million-year-old gravels from the south buried under basalts north of the present Grand Canyon (gravels that could not have been deposited after the western Grand Canyon was formed) reveal an uncompleted Canyon at that point. But only 5.3 million years ago small fossils from the Mancos Shale of the Colorado Plateau suddenly appeared in California's Salton Trough, testifying that this energetic young stream feeding the new Gulf of California had sliced through the scarp of the Grand Wash Cliffs to claw its way nearly halfway across the western Kaibab Plateau, where its steeper gradient and lower elevation redirected the old Colorado, thus stealing the thunder from the Kanab drainage and ultimately causing its reversal into the low-gradient, mild little creek that today flows in a very oversized canyon. By 3.8 million years ago, western Grand Canyon was only 300 feet less deep than it is today. By 1.2 million years ago, the Colorado had carved out its present configuration. Lucchitta estimates an astoundingly rapid downcutting of Grand Canyon at a rate of 0.015 to 0.038 inches per year—an inch every 40 years through solid rock!

Does Grand Canyon appear as if this was how it all happened? Not to me. Subjectively, when I look at the western Grand Canyon—particularly at lower Granite Gorge downstream from Diamond Creek—it looks older than Marble Canyon and much more than 5 million years old. Do we need yet another hypothesis?

Who knows? For inspiration I gazed across Surprise Valley. No answer down there. But there was water beyond it, which all five of us needed a lot worse than an answer. The last section of trail dropping westward from Surprise Valley zigzags over to Deer Spring, a smaller version of Thunder Spring that contributes about half the water to Deer Creek. The other half comes from Vaughn Spring, hidden a couple of miles farther up Deer Creek to our right.

Deer Creek is a paradise. Actually, most of it is simply a beautiful, clear, cold creek, rushing and gurgling, flanked by cottonwoods, in a canyon dotted with Anasazi ruins and strange walled terraces for agriculture and guarded by cliffs in which are hidden the Anasazi's ubiquitous stone granaries. A mile from the Colorado River, Deer Creek slices into the Tapeats Sandstone with the impossible narrowness of an undulating laser beam to create perhaps the most beautiful sandstone grotto on this planet.

At the upper end of Deer Creek Grotto, the creek rushes along a stone trench flanked by wide floors of Tapeats Sandstone and shaded by huge cottonwoods flickering emerald leaves in the sunlight. Each of these giants—living signs in the desert advertising "water here"—evapo-transpires hundreds of gallons of water daily in summer. A small price for their beauty. A bargain for the shade they offer.

Deer Creek fans across the sandstone floor, then plunges fifteen feet into the head of the grotto. From here Deer Creek alternately flows along sandstone beds and pours over abrupt falls, some merely a foot or two high, others fifteen. The grotto deepens so quickly and so impossibly narrowly that new visitors peering down from the narrow trail above (on the same level as the stone floor at its head) cannot imagine that a human could enter it to follow that twisting, roaring flume so far down below them. Or, once down there, could escape.

Even without a rope, a couple of routes exist downstream from the upper falls, but a rope makes more sense. Descending the Tapeats Sandstone is like Christmas morning. The last time I descended solo.

The last step is a leap into a pool below a giant chock-stone lodged in the creek bed, causing a waterfall. Common sense tells you that trying to walk downstream in the racing stream itself will lead to disaster, but it doesn't. Knowing this, I walked along little ledges adjacent to the creek when I could, but in the creek if no dry footing existed. To complicate matters, I avoided easy routes in the creek to traverse the rough strata of the Tapeats walls wherever any possible route existed, clinging with fingers and toes. Practice.

The grotto here offers the most photogenic sculpted sandstone I have seen. For uncounted millennia, obeying the laws of fluid physics and taking advantage of weaknesses in the adjacent strata of tan, striated Tapeats, Deer Creek has carved the bedrock into swirls, flumes, rooms, falls—all in smooth, organic curves that make one cry out in delight. Once I brought sculptor Jerome Kirk here. He was speechless for several seconds.

"Interesting?" I asked.

Kirk paused as if my question was obviously rhetorical, yet not, then admitted, "Oh, yes."

I continued downstream past more striated, curving walls echoing with the music of the creek. I hopped off ledges and boulders, waded deep pools, traversed cliff faces, and could not shake the feeling that I was the luckiest person in the universe. Finally I came to a vertical fall of maybe a dozen feet to the pool below, whose depth was anyone's guess. It looked deep, and both walls flanking the falls were undercut. Without a rope, the climb back up looked impossible.

Hammered into a fissure in the sandstone immediately above the drop was a climber's bolt. An old one. Rusty. Someone had descended here before. I looked down into the pool. The grotto beyond it, now a hundred feet below the trail, extended about a hundred yards before Deer Creek dived out of the Tapeats cliff into space for a hundred feet to form Deer Creek Falls, the most impressive waterfall visible from the Colorado River. Even though short and dangerous, this inaccessible slice of Deer Creek Grotto beckoned impersonally. No rope, no dice. I turned away. But someday I will return with a rope.

I climbed back out, then followed the path from upper Deer Creek Grotto to the Colorado River. The route bypasses the grotto by following two hundred yards of narrow ledge of Tapeats Sandstone level with the stone floor at its head. Only two feet wide and sprinkled with sand that acts like ball bearings under an ill-placed foot, this ribbon of trail is serious. Slipping off would kill you. By comparison, those deep gulches that Wile E. Coyote has dropped into while trying to outwit that witless cartoon roadrunner seem as wide as Grand Canyon. It is so narrow in one spot that you could jump across that hundred-foot drop. You could. But I've never seen anyone try.

This ledge too is an Anasazi route. Scattered under overhangs here are dozens of handprints—old handprints, created by taking an ocher pigment mixed with some liquid base, probably water, and blowing it, either from the mouth or through a hollow bone, over a hand held flat against the wall. Why handprints? Who knows. Prehistoric people the world over have left such prints on protected cliffs. Maybe they are here to mark a territory. But more likely kids made them in play. Most of the prints are child-sized. Either way, this place has been special for at least a thousand years.

I continued toward the river, abandoning the shade on sinuous bedrock, the flicker of emerald and diamond as the breeze rustled the cottonwoods, and the reassuring rush of the creek over polished gravel. The route seemed to end abruptly at the edge of a two-hundred-foot cliff above Deer Creek Falls. The view upstream from here (Mile 136) of the narrowest section of river in Grand Canyon, Granite Narrows, is fantastic, a scenic overload.

In the dark it's even worse. A few years earlier, on a flow of about 25,000 cfs and in full moonlight, Michael Boyle, Geoff Gourley, and I jumped into a friend's little inflatable powerboat (used by daylight for upstream travel to facilitate the environmental research described in the next chapter) and blasted like hell upstream from here into Granite Narrows. Our goal: navigate an upstream run of Tapeats Rapid two miles ahead. The challenge: navigate every other rapid between here and there. The result: frightening.

The river seemed to race past us. Flecks of reflected moonlight zipped past as streaks at impossible speed as we struggled upstream against the current through the stationary waves and holes. The boat's floor loaded with water and became too heavy for an up-run. We veered to shore, drained it, then raced up again. One rapid after another fell behind us. Each time we turned back to shore, drained the boat, then powered upstream again. It felt bizarre to me not only to be traveling upriver but also not to be in control of the boat. I had never been on the Colorado except as a boatman. Finally Tapeats Rapid roared ahead of us. Running up it looked impossible.

"If this thing goes over, somebody grab the motor!" our friend yelled over the crescendo of roars as he raced us up the eddy adjacent to the tail waves at top speed before crossing the eddy fence and challenging the rapid itself.

Sure, I thought to myself, we flip this boat in the dark in a maelstrom of water going the wrong direction and one of us is going to grab the motor. And then what? Sink, of course. No life jacket will hold a man plus a motor above the surface. "Okay, no problem."

The motor roared into red line at top rpms, but we no longer moved upstream. Tapeats Rapid flowed past beneath us as we revved to the max. No way. We stopped, drained the floor, and tried again. Still no way. Too much current. It was impossible.

In defeat, we spun downstream to navigate the enchanting moonlit stretch of Granite Narrows. Across from Deer Creek Falls we executed a final sharp turn.

Boyle fell out—heels over head at the top of Deer Creek Rapid.

But another extremely quick doughnut on the frigid moonlit water rescued him at the brink.

Sure. Grab the motor.

I'd like to spend a week here—or months, for that matter. This hike from Tapeats Creek to Thunder River to Surprise Valley to Deer Creek and the grotto is my favorite place in Grand Canyon. So much of it remains unexplored for me. For instance, instead of turning left from Tapeats to trek up Thunder, one can continue directly up Tapeats Creek, in the creek, between cliffs. Tapeats Spring, the ultimate source, is only about three hard miles beyond. Within Tapeats Spring is a cave, the kind of cave you can boat in for miles. Connie and I hiked there ten years ago but could not enter the cave. We had run out of time. I'm still waiting. It is a good thing to wait for.

12

There once was a canyon as beautiful as Grand Canyon, perhaps even more beautiful. True, it was only half as deep and less than two-thirds as long. And even though the same Colorado River flowed through it as here, this canyon held no rapids; its strata were soft sandstones, and the gradient of the river was only two feet per mile (compared to eight feet in Grand Canyon). Instead of rapids, it offered only a few riffles that a child could float in a life jacket. But this 170-mile-long canyon was beautiful enough to bring grown men to tears, to make career politicians regret most that they had ever cast a vote to destroy it, and to give birth to an organization of eco-defenders whose icon against the destruction of the West is a monkey wrench.

Wallace Stegner wrote, "Awe was never Glen Canyon's province. That is for Grand Canyon. Glen Canyon is for delight." Glen Canyon—once a fairyland of stone and flowing waters—is now used by the Bureau of Reclamation as a stock tank to store 27 million acre-feet of water.

Glen Canyon still exists, just like the Bering Land Bridge still connects Alaska with the USSR. But as a river canyon that we will ever see, Glen Canyon is gone, another victim of water development in the West. The citizens of the United States lost an irreplaceable treasure—and a potential national park—to feed many of those sacred cows. The logic behind the bureaucratic trade leading to its sacrifice makes me worry about the future of Grand Canyon.

Glen Canyon was dammed to make the desert bloom and bank accounts swell—in a reverse of Robin Hood's modus operandi, by taking from the taxpayers and giving to the . . . well, to the recipients of subsidized water and to the brokers of hydropower. The reasons for our losing of Glen Canyon stand as a lesson that we must still guard Grand Canyon. Glen Canyon Dam is significant not only because it augurs Grand Canyon's future but also because the impact of its operation may steal what Grand Canyon offers to our children before they ever see it.

Glen Canyon died because the Colorado River did not hold enough water to fulfill allotments promised by the 1922 compact discussed in chapter 5. Philip Fradkin notes that in 1958 Congressman Wayne Aspinall of Colorado warned Secretary of the Interior Stewart Udall of Arizona, a proponent of the Central Arizona Project, "Anyone who knows the river and the river's history knows that there is not sufficient water for the Central Arizona Project, as far as its continuing efficiency is concerned, unless one of two developments takes place. The first is that the Central Arizona Project takes water that rightfully and legally belongs to other basin states. The second is that the supply of the river basin is increased by some means or other."

Good sites for a dam that would be big enough to solve this problem are rare. After World War II the Bureau of Reclamation told Congress that the best were in Dinosaur National Monument (now a national park), where the Green and Yampa rivers join at Echo Park. The Park Service hired landscape architect Frederick Law Olmsted, Jr., to assess this idea. He concluded, "Construction of dams at these sites would adversely alter the dominant geological and wilderness qualities and the relatively minor archaeological and wildlife values of the Canyon Unit so that it would no longer possess national monument qualifications."

This report barely rippled the pond of pork-barrel politics. But the issue of the people's right to retain their National Park system was so basic that conservation organizations took a serious and, for the first time, united stand against these dams. David Brower, then executive director of the Sierra Club, joined forces with twenty-nine organizations in a solid front. With them stood the most improbable of allies. Worried that large dams on the upper tributaries of the Colorado might compromise flows and electrical revenues from Hoover Dam, the Los Angeles Department of Water and Power leapt into the arena to protest that the interest incurred in financing the dams in Dinosaur would be ridiculously exorbitant.

Brower (this was a decade before he would find and enlist his whiz-kid ally, Jeffrey Ingram, in the fight against the two dams in Grand Canyon) stood before Congress and tore the figures of justification published by the Bureau of Reclamation to shreds by correcting the Department of the Interior's bungled "ninth-grade" arithmetic. No, a dam in Echo Park would not conserve 165,000 acre-feet over alternative sites; it would save only 19,000 acre-feet, and so on. Although he was aided by several other prominent conservationists, it was Brower's testimony and analysis that killed these dams. But not the forces who wanted them.

The dams in Dinosaur were reincarnated in 1955 in a proposed trade-off to dam Glen Canyon. The bureau was uncharacteristically hesitant about this because the dam would have to be emplaced in Navajo Sandstone, a formation made up of weakly glued sand dunes of a desert larger than today's Sahara that covered most of the western United States 200 million years ago. Philip Fradkin explains:

> The dark red rock is actually solidified sand dunes and is
> rated moderately porous and highly absorptive. It extends
> above and below the 710-foot-high Glen Canyon Dam and

serves as the leaky, crumbly frame for the gleaming white structure. A panel of three engineers and a geologist said of the canyon walls in 1922 that, "There is a tendency for the rocks to fall off in blocks." One of the engineers later wrote: "It does not seem feasible to build any type of masonry dam of the necessary height for effective storage on the soft sandstone of Glen Canyon."

So why and how did the bureau build a 710-foot dam there? They changed their minds about the rock, for one thing. For another, President Eisenhower signed the $900 million Colorado River Storage Act in 1956 (which would cost far more and would subsidize some farms with five times more money than what those farms can sell for on the open market). Because this act was amended to prohibit any large storage structure from being built within a national park or monument, it sedated conservationists worried about Rainbow Bridge—the world's largest freestanding natural arch (4.8 miles up Aztec Creek from the Colorado in Glen Canyon and which was flooded anyway)—and prevented serious opposition to damming Glen Canyon. Hardly anyone knew that 2,000 Anasazi archaeological sites plus more than 200 historic sites would be submerged under Lake Powell.

A few curious people did float Glen Canyon after the appropriations but before the concrete was poured to see what they were about to lose. Most were shocked. David Brower was apoplectic. His remorse over his "sin" rings clear in his statement in Eliot Porter's book *The Place No One Knew*: "Glen Canyon died in 1963 and I was partly responsible for its needless death. So were you. Neither you nor I, nor anyone else, knew it well enough to insist that at all costs it should endure." He was so distraught, Marc Reisner notes, that "Brower's friends actually wondered whether he might shoot himself."

What makes a man so emotional over damming a canyon? A look at *The Place No One Knew* or C. Gregory Crampton's excellent *Ghosts of Glen Canyon* explains the emotion. When I look at these or talk to the old-timers who ran the river in Glen Canyon, I feel loss. Then anger. It's difficult to feel otherwise once you see what was lost.

But what about that crumbly, porous sandstone? Reclamation fixed it (temporarily). Before pouring the concrete they drilled hundreds of holes in the rock surrounding the emplacement (fifteen miles upstream from Lee's Ferry) then cemented steel bolts up to eighty feet long in them to hold the rock together. The last cement was poured in 1963, at which time storage began.

Although the bureau still tightens these bolts, about 2,500 gallons of water per minute seep and leak around the dam through the sandstone and its interfaces with the concrete. Allegedly this flow is increasing, though the bureau denies it. But this is nothing compared to what is happening in the 1,960 miles of shoreline along the canyon walls of 186 miles of the Colorado River, 70 miles of the San Juan, and many miles of other tributary canyons in this, the world's longest reservoir. These sandstones and shales absorb more than a million acre-feet per year. I have tried to imagine what this means. Are the walls of the canyon slowly turning to slurry? Will they catastrophically collapse into the reservoir? Will the Navajo Sandstone holding the dam calve away to leave a solitary cement wonder standing as an isolated butte in the center of a liberated Colorado?

The dam was designed not just as an arch dam but also to stand there as an island, as a gravity dam of 10 million tons of concrete—4.9 million cubic yards—as both a pyramid and an arch with such a substantial base that it would not budge even if the canyon walls around it vanished. Will it? Hard to say. I do know that I don't want to be downstream as we find out. It makes me nervous just thinking about it.

What was Glen Canyon really like, other than being one of the most remote regions in the United States? Was it *the* incredibly beautiful fairyland among river canyons in the West? The consensus is that it was. I could quote pages of descriptions of Glen Canyon and its lush tributary canyons that strain both the descriptive powers of the English language and the imagination of the reader. But I won't. Well, on second thought, maybe I will.

While the dam was being constructed, the late Edward Abbey and Ralph Newcomb each inflated a small, cheap vinyl boat and put in at Hite, 167 miles upstream from Lee's Ferry, to see Glen Canyon before such a journey would require a submarine. They tied their boats together and spent a week floating down the condemned canyon. Abbey, the future iconoclastic advocate for preservation of the wilderness of the Southwest, was carried by the Colorado into a true paradise, but a doomed paradise, he says, one to be sacrificed not to necessity but to avarice. In his classic book *Desert Solitaire,* Abbey describes his journey:

> We drift down the splendid river, deeper and deeper and deeper into the fantastic. The sandstone walls rise higher than ever before, a thousand, two thousand feet above the water, rounding off on top as half-domes and capitols, golden and glowing in the sunlight, a deep radiant red in the shade.
>
> Beyond those mighty forms we catch occasional glimpses of eroded remnants—tapering spires, balanced rocks on pillars, mushroom rocks, rocks shaped like hamburgers, rocks like piles of melted pies, arches, bridges, potholes, grottos, all the infinite variety of hill and hole and hollow to which sandstone lends itself. . . .
>
> Neither of us can believe that very soon the beauty we are passing through will be lost. Instinctively we expect a miracle: the dam will never be completed, they'll run out of cement or slide rules, the engineers will all be shipped to the upper Volta. Or if these fail some unknown hero with a rucksack of dynamite strapped to his back will descend into the bowels of the dam; there he will hide his high explosives where they'll do the most good, attach blasting caps to the lot and with angelic ingenuity link the caps to the official dam wiring system in such a way that when the time comes for the grand opening ceremony, when the President and the Secretary of the Interior and the governors of the Four-Corner states are in full regalia assembled, the button which the president

pushes will ignite the loveliest explosion ever seen by man, reducing the great dam to a heap of rubble in the path of the river. The splendid new rapids thus created we will name Floyd E. Dominy Falls, in honor of the chief of the Reclamation Bureau; a more suitable memorial could hardly be devised for such an esteemed and loyal public servant.

Abbey's vision of eco-sabotage to rescue this priceless region gnawed at him. A few years after the first edition of *Desert Solitaire* appeared, Abbey expanded his vision to reveal how this rescue might be pulled off. Abbey's unknown hero solidified into a Vietnam veteran named George Hayduke, who discovers he is a misfit among the domesticated herds of humanity he finds grazing placidly in front of their TVs across North America from sea to shining sea. Hayduke loses himself in Canyon Country. Ultimately, on the run from various establishment baddies, Hayduke meets three other kindred spirits. Yes, they are all disgusted beyond tolerance by the destruction they witness in the Southwest, and they decide to strike back. They become a four-person American Revolution: the Monkey Wrench Gang. Abbey opens his book with a now classic scene in environmentalist literature, the detonation of Glen Canyon Bridge during its dedication ceremony as a teaser for the ultimate sabotage of the nearest large structure to it. Abbey's *Monkey Wrench Gang* compels so many real kindred spirits that it has become an almost sacred text. It inspired the hardcore environmental group Earth First!—which uses a monkey wrench as an icon. Some of the Bureau of Reclamation boys go into tachycardia at the sight of one.

Was Ed Abbey's prose on Glen Canyon merely good copy calculated to land him on the literary map? Abbey admitted to editor Lyman Hafen, "Always I was drawn back to the mysterious world of Glen Canyon, the living heart—so it seemed to me then and so it seems to me now—of the entire Colorado Plateau province. Glen Canyon was the most beautiful place I have ever seen."

Any conclusion that only hard-core environmentalists mourn

186

and resent the fate of Glen Canyon is wrong. An army of conservatives would like to get it back. For instance, W. E. Garrett (writing in the *National Geographic*) reported a change of heart by veteran Arizona senator Barry Goldwater, writing that "of all of his Senate votes, the one he most regrets was the one for Glen Canyon Dam."

Aside from remorse and fantasies for rescuing Glen Canyon, the most important question is, What is Glen Canyon Dam doing to Grand Canyon? Although Glen Canyon Dam put off Hoover Dam's fate of being choked in silt (by 1963 Lake Mead's capacity had dropped from 32,471,000 acre-feet to 30,755,000 due to siltation), Glen Canyon is substituting itself as a sacrifice. Silt collects in Glen Canyon at about 35,000 acre-feet per year—at the cost of beaches in Grand Canyon. (The reservoir also loses about 450,000 acre-feet of water annually from its 161,390-acre surface through evaporation.) At this rate of siltation, assuming that the contribution from overgrazed ranges in the Navajo country does not increase (already Navajo land supplies only 2 percent of the water but 37.5 percent of the silt load to the river), Glen Canyon Reservoir will be filled to the brim with mud in 770 years (official estimates range from 316 to 1,000 years). Of course, most of its utility will have vanished long before that distant time, and desilting it would cost more than building a solar power plant or a nuclear plant plus a waste disposal facility unless—and this is the only real solution—the bureau builds a slurry pump to siphon Glen Canyon sediment into Grand Canyon for a month each year. Ultimately, Glen Canyon Dam will be an expensive sacrifice for ourselves and our posterity—especially our posterity. Marc Reisner summed it up:

> What federal water development has amounted to, in the end, is a uniquely productive, creative vandalism. . . . a vandalism of both our natural heritage and our economic future, and the reckoning has not even begun. Thus far nature has paid the highest price. Glen Canyon is gone. . . . [But] the real vandalism [is] the financial vandalism of the future.

. . . Who is going to pay to rescue salt-poisoned land? To
dredge trillions of tons of silt out of the expiring reservoirs?
To bring more water to whole regions, whole states, dependent
on aquifers that have been recklessly mined? To restore
wetlands and wild rivers and other natural features of the
landscape that have been obliterated, now that more and more
people are discovering that life is impoverished without
them? We won't have to. Our children probably won't have to.
But somewhere down the line our descendants are going to
inherit a bill for all this vaunted success. . . . It will be a
miracle if they can pay it.

I cannot help but wonder too, is this how we should spend
our hard-earned tax money—$70 million on one more of 355 bu-
reau dams? (We could learn what citizens really want by per-
suading the IRS to place on income tax forms a new section below where we sign: "Please check your top five priorities for
government spending to receive the taxes you just paid.")

Be that as it may, between 1956 and 1963 the Bureau of Reclamation—allied with powerful agricultural interests—raided
and stole from us, the American citizenry, one of the most beautiful national parks on the continent. It's as dead now as the eighteen workers killed while building the dam. Today, after signing
into law the National Environmental Policy Act (NEPA) in 1970,
which requires that "the federal government use all practicable
means to create and maintain conditions under which man and
nature can exist in productive harmony," Glen Canyon would
surely be a national park—were it not dammed. In fact, NEPA
emerged partly *because* Glen Canyon had been dammed. Still unclear, though, is how Glen Canyon Dam continues to alter the
Colorado and degrade the river corridor in Grand Canyon.

Of course, some effects are obvious. Except after storms have
flooded its tributaries, the river is no longer the warm, silty Rio
Colorado that once flooded every June and trickled in December.
Today it is a chilly, translucent green flow that rises and falls daily
depending on the electrical needs of Phoenix (the customer for

about half of the hydropower from Glen Canyon Dam, which can put out a maximum of 1.352 million kilowatt hours). Annually Glen Canyon Dam cranks out 407.4 million kilowatt hours sold through the Western Area Power Administration (WAPA) to member brokers of the Colorado River Electrical Distribution Association (CREDA) for only a quarter to half of the normal market rate, which then sell it to seventy-six utility districts. As a by-product, nutrients and sand the river used to deposit within the river corridor as it ebbed from seasonal floods today remain trapped in Lake Powell.

It may seem that a shortage of sand would be the least of Grand Canyon's problems. Sand still finds its way into food, clothing, bed rolls, cameras, and every orifice of 21,000 boaters annually. But this is because we camp on beaches. Camping on boulders is hell. The problem is that with Colorado sediments from the Canyon being choked off, its beaches are shrinking drastically in size and number. How fast is a question now being heavily researched, but now during peak seasons we occasionally must camp on heaps of boulders where once there was a beach.

The Bureau of Reclamation is responsible for yearly releases of water from Glen Canyon, and it sets releases monthly. WAPA determines the daily and hourly fluctuations within those limits. This arrangement leaves the bureau's men in the dam monitoring the hourly releases—but not setting them. WAPA men in Montrose, Colorado, mastermind fluctuations with a regard for nothing but the bottom line, and the people standing to make the most are the power brokers in CREDA and WAPA.

Because the river fluctuates daily, even hourly, by a drastic 500 percent in sharp ramping rates due to manipulations designed to generate more money from hydropower, the remaining sand (and camps) in the Canyon are accelerating toward Lake Mead. Mooring boats at night requires expert planning and care or they will end up stranded on shore or abraded or ripped against rocks that emerge with the fluctuating levels. Further, we boatmen must take our passengers through rapids at more dangerous levels than otherwise. In short, the wild Colorado in Grand Canyon and all that it created and supported are vanishing or are already gone. Contrary to some claims, Glen Canyon Dam did not make whitewater boating possible. The river was always navigable, even on peak flows in June.

The Colorado River itself has changed and its habitat has been modified. Because it no longer fluctuates seasonally, dirty beaches stay dirty; little new driftwood enters the Canyon; tamarisks invade and grow into impenetrable thickets that monopolize existing beaches; and new rapids such as Crystal, Granite, 24 Mile, Fossil, Specter, and Bedrock remain nearly as nasty as the tributary canyons made them. Before the dam, in summer the Colorado's water reached 80° F and was silty. Several species of native fish, species that the National Park Organic Act of 1916 is supposed to protect, are becoming rare or have already become extinct (as in the case of the endangered Colorado Squawfish and the bonytail chub) because the new Colorado is too cold for breeding and because exotic species have been added. And people, by the way, who are tossed into this new arctic river during whitewater accidents become hypothermic and find their lives much more seriously threatened than before.

The power of money thus influences the quality of the river experience for everyone. If challenged in court, some of these effects would no doubt be judged illegal modifications of existing natural conditions in a national park, which is proscribed by the National Park Organic Act. So why does WAPA release water the way they do? Because of the money politics among brokers of hydropower in the West and also the convenient lack of cost involved in changing the rate of hydropower generation. Nuclear, oil, and coal-fired plants cost money to fire up and cool down; dams do not. Hydropower revenues are important too because they help repay the cost of building and operating Glen Canyon Dam and of constructing associated bureau irrigation projects in the Colorado River Basin.

In a federal government gone wild with self-spending to

the point of accruing a $3 trillion debt, these are powerful factors. Even so, public concern over upgrading the generators in Glen Canyon Dam from 1,150 megawatts to 1,352 megawatts—thus increasing peak flow releases by 4.8 to 14.6 percent and creating even more drastic fluctuations for peaking-power generation—finally forced Interior Secretary James Watt to accede to National Environmental Protection Act requirements in late 1982 by initiating Glen Canyon Environmental Studies (GCES), a multimillion-dollar investigation of the impact of fluctuating flows and floods on Grand Canyon. Interestingly, Glen Canyon Dam was not built primarily to maximize hydropower by fluctuating flow releases. Its primary function and sole justification was to store enough water to deliver 8.23 million acre-feet yearly to the lower basin states. The fluctuating flows result solely from brokers lobbying to buy the cheapest kilowatt hours in the West and reap a higher cut from consumers.

The bureau assigned coordination of GCES research to Dave Wegner, a bureau biologist with master's degrees in civil engineering and biology. By 1991 Wegner was coordinating 140 researchers from the Bureau of Reclamation, the U.S. Geological Survey, Arizona Game and Fish, the National Park Service, the U.S. Fish and Wildlife Service, plus private consultants. They were working on forty investigations of the impact of the dam's patterns of release in the areas of sedimentology, hydrology, fish biology, beach erosion, plant and bird ecology in the riparian (streamside) corridor, and recreational impacts. As a by-product of all this, Wegner fell under the spell of Grand Canyon.

On December 1, 1985, rangers Ruthie Stoner, Mark O'-Neill, and I (I was moonlighting), plus Dave Wegner put in our four boats at Lee's Ferry to pick up observers stationed earlier at Hance, Crystal, and Lava Falls rapids to observe the trouble boaters faced at different flows. During the fall, boaters were so rare that these observers saw only one party weekly instead of the five per day of summer. But because the summers of 1984 and 1985 had been record years of higher than average runoff (20.8

and 19.1 million acre-feet), observations of low releases had had to wait until now.

"This seems ridiculous," I said to Wegner, playing the devil's advocate.

"I know, I know," he admitted in frustration. "But summer would not have given us data on fluctuations. And now that they're down there, we have to pick them up." (Using a helicopter to pick them up would have rubbed the "anti–air traffic" Park Service's fur the wrong way.)

But for Wegner this was not just a rescue mission; it was an on-site inspection of one of "his" investigations (actually, despite the bureau's picking up the tab for this trip, the project was organized by Martha Hahn of the Park Service). It was also a chance for him to become more familiar with the Canyon. The trend in the assessments was becoming so obvious that Wegner was finding himself between a rock and a hard place.

Some of the effects of the dam's peaking-power release patterns I listed above were more severe than most of us realized. For example, the bureau's parsimony with storage in the reservoir, followed by sudden flood releases—the massive flood of 97,200 cfs during June 1983 and the subsequent floods of 1984 and 1985—had seriously eroded many remaining beaches, creating a huge net loss above Mile 120, and had scraped away the Canyon's riparian community (itself an artificial product of dam releases). They had also eliminated 95 percent of the marshes in Grand Canyon. Worse, WAPA's daily fluctuations were also more damaging to nearly every other resource examined in the investigations than constant flows appeared to be. These might have been dry statistics to any supervisor who had not experienced the river, but already Grand Canyon had impressed Wegner as the experience of a lifetime.

"This is such a tremendous resource," he volunteered to me, only to lapse into private thought. Was he thinking that the data from the first $7 million of research proved that bureau management required an overhaul? The rub was this: Glen Canyon

Dam's primary legal function—set by an imbroglio of legal compacts, treaties, and acts following the 1922 compact all combined into what is now commonly referred to as the Law of the River—is simply to store and deliver 8.23 million acre-feet of water to the lower basin states. Producing hydropower is secondary. But the reality is that the water Glen Canyon Dam delivers does earn money when it flows through the turbines, and while no one in WAPA or the bureau earned more money personally by releasing water in fluctuating peaking-power cycles, the buyers and sellers of hydropower through CREDA did; and everyone in the bureau and WAPA was under pressure to maximize revenues and lower costs. An additional complication is that (like the many generals retiring from the armed forces to become vice presidents in defense industries) bureau men "retired" to become CREDA executives. As a result, the inescapable conclusion that stable releases are environmentally more sound would be unpopular with Wegner's supervisors because stable flows cut revenues (by a mere tenth of a cent per kilowatt hour) even though they generate the same total electrical power (equivalent to 6.7 million barrels of oil per year). Wegner had integrity, but in the bureau, integrity took a back seat to money. To protect Grand Canyon, the bureau and WAPA had to modify releases to a stable pattern similar to nature's own, but doing so would cost the bureau millions because revenues would be replaced by somewhat more expensive power from nonbureau generating stations. Just how expensive would this be?

The economic cost of Glen Canyon releasing only stable "baseload" flows is the vital question to those for whom money is the only bottom line. Estimates range wildly. Even WAPA's own unimaginative estimates for 1991 range from $21.6 million to $39.6 million, and for 1995 the range would be $20.1 to $32.6 million. A more sophisticated mid-1990 estimate by the Environmental Defense Fund's ELFIN Model indicates potential costs of $8.5 to $19 million in 1991 and $6.8 to $19 million in 1995. The Environmental Defense Fund's 1990 *Estimates of Economic*

Impacts report to the House of Representatives notes that this change is "significantly less than 1% of the overall power system costs." The most recent (and unpublished) estimate boils it down to the cost levied on the average consumer. If Glen Canyon went to stable flows, the average household would pay less than a penny more daily for electricity—twenty-five cents per month. Is Grand Canyon worth twenty-five cents?

The final GCES summary report was completed in January 1988. The National Academy of Sciences gave it fair technical marks but was critical of the study's planning and overall management (which had been constrained by the initial research design and inherent government bureaucracy). Tellingly, the GCES report's introduction stated:

> *This study was not intended nor designed to lead directly to changes in dam operations.* Any decision to make operational changes would require feasibility studies and National Environmental Policy Act (NEPA) compliance activities to assess the impact of those changes on the primary mandate of the Colorado River Storage Project (water storage and delivery and power generation), as well as on the environment and recreation.

In short, any mandatory change in dam operations would be spurred by an environmental impact statement.

Despite the inference that no one really had to act on the contents of the report, the report itself concluded in no uncertain terms that current flow regimes from Glen Canyon Dam are damaging resources unnecessarily within Grand Canyon National Park. The report then identified changes in operations— evaluating four potential alternative regimes—that would reduce those impacts and still meet the legal downstream releases from the dam in a more steady state. No regime included peaking-power releases.

But because some projects had not been integrated with one another in advance by design, the National Academy of Sciences was critical of the study and suggested that the Bureau of Reclamation repeat some of the research—as a Phase II program—scientifically directed by Duncan Patten of Arizona State University, a resource management Ph.D. scientist. Then—prompted by urging from the Grand Canyon Trust, the National Wildlife Federation, and a grass-roots letter-writing campaign—on July 27, 1989, Secretary of the Interior Manuel Lujan decided to shift Phase II to a full environmental impact statement to integrate research projects to delineate the effects of flow rates and regimes on all resources and "to assess the need for measures to minimize the impact of Glen Canyon Dam on the downstream environment and ecological resources" of Grand Canyon National Park.

What will this mean for the Canyon in the long run? The GCES project was an excellent beginning. Representative George Miller and Senators John McCain and Jay Rhodes have independently introduced legislation to direct the interior secretary to implement constraints on operations of Glen Canyon Dam now, even before the environmental impact statement is completed. But one lesson here is paramount: Without public pressure, nothing would have changed, and without continued public pressure, the Canyon will not be saved. Tellingly, Wegner wrote to me that "While they may get rid of me, I hope that you do not stop writing about the [impact] that has occurred and will continue to occur to the resources of the Canyon."

A shocking example of how bad federal management can be is provided by John Skow in a recent article entitled "Forest Service Follies." Skow describes the federal investment of $62 million per year to cut 400-year-old Sitka spruces in Alaska's Tongass National Forest to sell to Japan for a mere two or three dollars per tree at an unbelievable loss. Skow writes: "In 1983 the [forest] service lost 91 cents on every dollar invested in Southwest Alaska—if invested is the right word. In '84 it lost 93 cents on each dollar. In '85 and '86, 99 cents." A major priority of this deficit spending—and destruction of this rare primeval forest—was simply to keep the logging communities economically viable. Skow notes that it would have been cheaper to mail the 1,400 lumberjacks checks for $36,000 each, send them on a cruise, or something—and tack on instructions to leave the forest alone. To do so would not only save the forest but would save money too. Such monumental and needless destruction of our natural resources—coupled with a scandalous waste of tax dollars—by our government persists year after year because we citizens wallow in complacency.

We who live in the Canyon, floating the river seemingly forever, consider these issues serious and immediate. We know the drastic extent of the current damage and that which will occur if nothing changes upstream. James Bishop quotes boatman John Parsons's focus on the big picture: "The Bureau of Reclamation has done more damage than Edward Abbey's Monkey Wrench Gang ever dreamed of doing."

Grand Canyon teaches the lesson that we ought to take nothing on this planet of ours for granted. There may be billions of other terrestrial-type planets in the universe, thousands of which may be habitable—but this is the only one that is convenient. In *The Home Planet*, K. W. Kelly quotes astronaut Loren Acton as saying, "When you look out the other way toward the stars you realize it's an awful long way to the next watering hole." Hence, until someone invents a practical hyperspace drive, it would be prudent to treat Earth as the most precious jewel in the universe. It is.

The level of frustration involved in waiting for ten years of research in hopes that it will vindicate what we already know is revealed by Ed Abbey's practice of spinning off plans to blow Glen Canyon Dam to smithereens. Abbey took a breather in his direct-action approach, however, to point out that, even if his plans—and all our plans—fail,

> if we don't do it, nature will. In a few more centuries the dams
> will be filled with silt and mud, and will become great water-
> falls. Then, as erosion does its work, the dams will be reduced

to polished stumps of concrete and re-bar, forming rapids full of V-waves and suck holes—a challenge to boat people, nothing more. Any river with the power to carve through the ancient limestones, sandstones, granite, and schists of the Kaibab Plateau will have little trouble with the spongy cement deposited, once upon a time, by some dimly remembered clan of ant folk known as the Bureau of Reclamation.

Maybe. Again it is more likely that when Glen Canyon Dam silts up the Colorado will not destroy Glen Canyon Dam but will slice through the sandstone and detour around the dam to leave a gleaming white megalith over which archaeologists of the future would theorize. I'm tempted to leave this book in a time capsule nearby. Maybe I'll drop it off when I take Cliff and Crystal to Page to the scenic overlook above the dam that created North America's longest reservoir.

13
The Honeymooners

Under a ribbon of stars winking from the center of our galaxy, the river rushed past, shoving and jostling the boats like a rude commuter in a subway terminal. The boats were safe in this micro-eddy roiling against the cliff face at Mile 152. We had tied them to shore in five bomb-proof locations and had anchored them with aluminum chocks and carabiners and a spider web of nylon ropes and straps to float them as one unit free of the jagged rock.

At intervals now the river surged in a boil next to my boat, sounding exactly like a hippo emerging. It was hard for me to ignore, too ingrained a habit. I had rowed and paddled fifteen hundred miles of African rivers filled with hippos and crocodiles, always in inflatable boats. One of the talents vital to surviving such rivers is an acute fear of sharp teeth in close proximity. I had been charged by both hippos and crocs who got close enough to me so that I could have patted their noses from the nonexistent security of my inflatable boat. I had not patted them, but I had spent many long hours swatting tsetse flies and patching boats ripped and shredded by their tusks and teeth.

Anyway, the roiling of this angry eddy triggered my conditioned neurons involuntarily. I found it impossible to relax and ignore the repeated hippolike "Whoosh!" so I lay on my sleeping bag on the rear deck and slipped Sony headphones over my ears and punched the play button. Dire Straits. The stars twinkled again between the narrow cliffs as side two of *Brothers in Arms* started. Weird horns and drums forced the phantom leviathan back under the frigid Colorado. Mark Knopfler's electric guitar and gravelly voice boomed in my auditory canals: "I'm a soldier of freedom. . . . "

I slipped into my sleeping bag and watched the river swiftly pass as if it were being sucked from the Canyon by a Gulf of California dying of thirst (which indeed it now is). Grand Canyon is a place of romance, of challenges larger than life, of forbidding landscapes beyond description, and of experiences as private as a first love. The romance of the Canyon pummeled me now. This place does something to people. It plucks atavistic neurons as if it were playing a flamenco guitar. And some dance until they die.

One effect of this river that I never would have guessed is that of psychoanalysis—it reveals the depth, or lack of it, of a relationship. Couples arrive at Lee's Ferry either with a thriving relationship or one held together by the weak glue of habit, or finances, or status, or insecurity, or you name it. The river will weld the people who are bonded in a healthy relationship even tighter—happy and dreaming together and newly aware of how blessed they are to have one another. In contrast, it will dissolve weak relationships like a lump of sugar.

Why? Because the Grand Canyon challenges people. Day after day it freezes them, scorches them, wears them out, torments them with thirst and hunger, drenches them repeatedly, hurries them up and then slows them to a dead stop, scares the living daylights out of them and then soothes and consoles them like some gigantic guru. It forces people to trek in blistering heat, climb on rock hot enough to refry beans and on exposed routes above falls that if they fell would mangle them like doomed bison driven off a cliff by Paleo-Indians. The Canyon sucks the moisture out of people fast enough to make a prune farmer turn green with envy. Then it showers them in a cold rain that will not stop. It strips away all sophistry. It makes people feel as insignificant in the scheme of things as a gnat and, worse, pointedly shouts that their lifespans here on Earth are just as fleeting.

People here fall out of love—and into love. The Canyon is the elemental catalyst of primeval passions, good and bad. It nurtures what is strong and destroys what is weak, especially in marriages. We see it all the time. After this trip, for instance, four out of our five married couples would split up before a year had passed. If that seems bad, consider the fate of the Canyon's most famous honeymooners, a fate that still tantalizes and spooks every couple who launches a one-boat trip down this river.

In 1927 a shy young woman named Bessie Haley, who had come to San Francisco a year earlier from Parkersburg, West Virginia, to refine her passion for drawing at the California School of Fine Arts, embarked with a girlfriend on an overnight ship to Los Angeles. Although only twenty-one, Bessie had already been through some rough times. Back east she had secretly married Earl Helmick and had lived with him for about two months, gotten pregnant, and then vanished to the West, where he sent her money for an "operation."

Fate had placed on this same ship a tall young adventurer named Glen Hyde. Hyde, the twenty-nine-year-old son of an Idaho rancher, had a passion for rivers. He had floated rivers in British Columbia and Idaho—including the Salmon—when such trips were anything but routine.

Electricity must have flowed between Glen and Bessie. On that short voyage down the Pacific Coast to the City of Angels, their destiny became the stuff of a Hollywood fantasy.

Bessie married Glen several months later in Idaho on April 12, 1928—one day after her Nevada divorce from Earl Helmick. Glen had a dream: to spend their honeymoon as the first husband

200

and wife team to run the Colorado River through Grand Canyon. Bessie wrote poetry and was an A student in drawing. Glen's plan was to set a record for speed and run every rapid on the river, then to produce a successful book that would launch them both on the lecture circuit.

In October 1928 they arrived at Green River, Utah, where Glen built a flat-bottomed scow twenty feet long by five wide by three deep. He equipped it with massive sweep oars and furnished it with bedsprings, a mattress, and a stove consisting of a sand box saturated with kerosene. Glen's river version of a Winnebago held everything they needed to be free from camping on shore. But despite its cozy appointments, Harry Howland, who had run some of the river earlier, looked at it and said something like, "It looks like a floating coffin."

On October 20 the honeymooners set off down the Green River. Bessie, only five feet tall and weighing only ninety pounds, could barely manage the long bow sweep that Glen had devised. Once, while they were rowing in Canyonlands, it swept her off the scow into the river. Glen wielded the stern sweep. Amazingly, they arrived unscathed at Phantom Ranch 26 days and 424 miles later, a record for that time, having floated past Lee's Ferry on about 14,000 cfs.

In Sockdolager Rapid, Glen's sweep knocked him in the chin and catapulted him from the scow. Bessie tossed Glen a rope and grabbed the flailing sweeps. This had been more terrifying than usual because the river was cold now and because Glen had refused to bring along life jackets.

At Grand Canyon Village the couple was treated as royalty. Emery Kolb (by now a veteran of two Canyon voyages and a permanent resident of the village) pressed the couple about life jackets and offered them his own. But Glen refused, saying they did not need "artificial aids." This has always made me wonder what Glen classified his scow as, a natural raft? He also refused to buy a couple of inner tubes for flotation from Fred Harvey's garage.

Kolb tried his best, but Glen said, "I'm going to do it without [life jackets] or else."

After hot baths and luxurious food, Bessie reportedly felt misgivings about their journey. Emery offered them his hospitality for the winter. The river corridor sat in nearly constant shade at this time of year, a cold fact that Emery and his brother, Ellsworth, had learned the hard way seventeen years earlier. But Glen would not even consider staying. And while nothing in Bessie's letters from the rim mentioned her fear, Glen himself wrote, "I would quit the river here, not on my own account, but from what they tell us, we're past the worst water." Tellingly, before the couple hiked back down Bright Angel Trail, Bessie spotted a new pair of shoes belonging to Emery's daughter, Edith, and said, "I wonder if I shall ever wear pretty shoes again."

Glen was obdurate, Bessie subdued. With photographer Adolph Sutro as a temporary passenger, they shoved off from Phantom Ranch on November 18, beaching at Hermit Rapid (Mile 95) to rendezvous with two mules loaded with supplies that Sutro had paid for. Several residents of the rim had hiked down to see them off. In 1973 Emery Kolb told O'Connor Dale that one of the men present saw Glen physically force Bessie back into their scow.

Sutro was highly critical of Hyde. He remarked to Dock Marston, "It was the most inadequately prepared expedition I had ever seen." Sutro characterized Glen as "decent, not domineering, but irresponsible" and self-important to the point of having "delusions of grandeur." His boating skills were negligible. Sutro also discovered that Glen had no money and nothing to eat, hence Sutro's gift of food. Revealingly, Sutro told Marston that, during those half dozen river miles, "Glen kept talking about the money" that would result from their vaudeville stint after the trip. This trip, in fact, appeared to be strictly a stunt undertaken for money.

An eyewitness at Hermit Rapid noted that after Glen forced

Bessie back into the scow, her face "registered stark terror." The pair then disappeared alone down the sunless stretch of river in First Granite Gorge. Disappeared forever.

Glen and Bessie missed their rendezvous with Glen's father, R. C. Hyde, at Needles, California, on December 9. A week or so later the elder Hyde commissioned a search plane, a Ford Tri-motor piloted by army lieutenants L. G. Plummer and H. G. Adams. Risking their necks by flying close above the river, on December 20 they spotted the scow floating in a quiet eddy near Mile 236 Rapid at Gneiss Canyon. The next day Emery Kolb hopped on the plane to identify the scow. It was *the* scow. But no plane could land near this section of Lower Granite Gorge, and the closest good boat was 148 miles upstream.

The elder Hyde asked Kolb to run down the river and investigate the fate of his only son. Emery wired Ellsworth in Los Angeles. At the mouth of Diamond Creek (Mile 226) they quickly reconstructed an old boat abandoned by a prospector. Both Kolbs, along with Chief Ranger Jim Brooks, headed downstream. Lining Mile 232 Rapid turned out to be tough—prompting them to suspect that this may have been the straw that broke iron man Glen's back.

Five miles downstream, they found Glen and Bessie's scow self-moored, with its bowline wedged between submerged rocks. Everything was shipshape: the bed, the stove, even Glen's rifle and coat and Bessie's box camera and diary. The honeymooners had carved forty-two notches in the gunwale to mark the number of days from Green River. This places the last notch on November 30. Emery noted that the scow was double-hulled and that the bilge space between the floor and the hull was swamped with enough water to have made the scow sluggish to pilot and difficult to moor or line while in the current. So, while the boat had not flipped, it had somehow lost its crew. The Kolbs and Brooks collected what possessions they could, then Emery severed the bowline holding the scow. (Later O'Connor Dale told me that in 1973 "Emery seemed real sad" when he admitted that he regretted cutting the boat loose, because the honeymooners may have been still in the Canyon somewhere and would have needed it.)

How had Glen and Bessie been lost? Only a few tantalizing clues to this mystery have been found. Bessie's diary was a disappointingly skimpy document, with circles marked for easy rapids, slashes for hard ones, and horizontal dashes for quiet stretches. It provided no detail about how she felt about the trip or their experiences. Her diary (long since destroyed by her family after being transcribed by Kolb) did place them three days downstream from Lava Falls (Mile 179.5) against heavy winds and low water by November 30.

Hualapai Indians hired by R. C. Hyde found footprints and an empty jar of lima beans at Mile 210, and three years later someone found Bessie's and Glen's names penciled on a board in an old, abandoned blacksmith's shop at Diamond Creek, with the date "November 31 [sic], 1928." By this time the Colorado was flowing at about 5,000 cfs—a nasty level for Mile 232 Rapid, with pinnacle fang rocks sieving the main flow. It seemed that whatever had happened to them had occurred somewhere along the first eleven miles below Diamond Creek, and if those notches and Bessie's diary were accurate, it had occurred on November 30 or December 1.

The Kolb brothers guessed that while Glen was scouting Mile 232 Rapid, either the bowline had slipped from Bessie's grip or the heavy scow had dragged her into the river and Glen had leapt in to help her. Perhaps, instead, the scow had wrapped or was pinned against the cliff face or fang rocks while running the rapid and had flipped them out. Either way, without life jackets their chance of survival in the cold water would have been minuscule. The only fact of which people were certain was that Glen and Bessie Hyde had vanished.

However it happened, Bessie Hyde had become the first

woman to float the 560 miles of the Colorado between Green River, Utah, and Diamond Creek.

But the story is far from over, and the mystery anything but solved. (I owe a debt for many of the following details to river sleuth Marty Anderson, geologist George Billingsley, boatmen O'Connor Dale and Regan Dale, and fellow writer-boatman Scott Thybony.) Forty-three years later, in the fall of 1971, on a twenty-one-day rowing trip run by Grand Canyon Expeditions, Bessie Hyde rose from the dead.

On nearly the last night, trip leader Rick Petrillo regaled his passengers around the campfire with the romantic mystery of the honeymooning Hydes. In a turnabout, Petrillo, Regan Dale, O'Connor Dale, and George Billingsley were themselves startled by one of their commercial passengers. Boatmen don't startle easily, and when they do, they take pains to avoid revealing it. But on this occasion, after Petrillo finished his tale an older woman named Liz admitted to O'Connor and George, matter-of-factly, that she was Bessie Hyde.

"Well, what happened to Glen?" one of them asked, playing along.

"He was a son of a bitch who beat me all the time," Elizabeth asserted. She explained that she had wanted to quit the river at Phantom, but Glen had forced her into the boat. She knew from Kolb that the worst rapids, Separation Rapid (Mile 240) and Lava Cliff Rapid (Mile 246), were downstream from Diamond Creek and that Diamond represented her last chance to escape. They had pulled into shore a few miles above Diamond, she continued, where she stabbed him while he slept, pushed his body into the river, shoved the scow off with everything aboard, then hiked out to Route 66, where she flagged down a bus. Later Liz pulled out a small penknife to show O'Connor, and with a twinkle in her eye, said, "This is what I did Glen in with."

Because O'Connor and Regan reckoned that Liz would have

had to stab Glen at least a hundred times with her little penknife, and because she seemed to be pulling their leg besides, neither took her seriously. Billingsley told me he too was unconvinced but later found himself wondering. "You never know," he admitted. But Regan told me she was "a spunky little old lady" who, with her story, "caught one of the boys and was reeling him in." O'Connor told me he jokingly called Liz "Bessie" a few times before the trip ended and got a mischievous grin out of her each time.

A joke? Maybe, but Liz's story is not devoid of support. For instance, while the Kolbs and Brooks were working on their rescue boat at Diamond, the elder Hyde had hiked upstream on the south side. A few miles up he had found footprints on a beach, which, because the water was dropping, would have lasted a long time in the clay-encrusted sand. And although the details concerning the prints are unclear, that camp and footprints fit Liz's story. Even more suspicious, Liz "confessed" at a camp a few miles above Diamond, virtually the scene of her alleged crime.

Was Liz really Bessie Hyde? Intrigued by this mystery, in 1985 Scott Thybony contacted Liz at her home in Ohio, which was only thirty or forty miles from Bessie's hometown. Liz remembered some details about the 1971 river trip—for instance, a broken arm a passenger suffered while they were running Horn Creek Rapid. But, Liz told Scott, she had no memory of a Glen and Bessie Hyde story, that the name Hyde rang no bell at all. And, by the way, she was not Bessie Hyde.

If Liz was joking, why was she? For weeks Liz had been a real trooper during hardships that included a snowstorm, equipment loss, and injuries. She had been assisted in her day-to-day camping routine by O'Connor's father, Eben Dale, who considered her a delightful little old lady but who had hiked off the trip at Havasu a few days before Liz announced she was Bessie Hyde. O'Connor thinks that, with her story, she was just looking for some attention. The trip had become boring for her after Eben Dale had hiked out.

River sleuth Marty Anderson became obsessed with this mystery, so much so that he went beyond what Scott Thybony had done in search of the facts. Like Thybony, Anderson admitted to me that he initially believed that Liz was Bessie but was suspicious because the tale was too romantic. He found a fellow passenger of Liz's who had kept a journal. In it he described Liz as a "tease" who "gave as good as she got" and "an aging Cleopatra . . . giving O'Connor orders."

Marty, too, phoned Liz. Yes, of course she remembered her Colorado River trip, but (again) she had no idea what Marty was talking about with this Hyde story: "I don't even know who Bessie Hyde is!" Anderson located birth certificates for both Bessie Haley and Liz (using her maiden name)—Liz was four years younger. A late 1920s newspaper notes that Liz attended a large family reunion while Bessie Haley was out West. Finally, while Bessie's passion was drawing, Liz was indifferent to art. Anderson's conclusion? Liz is not Bessie Hyde. So what really happened to Glen and Bessie?

Emery Kolb lived on the South Rim into his nineties, dying in late 1976. Upon clearing out his possessions of half a century, someone found a canoe-type boat hoisted up in his garage. In that boat next to a bundle of clothes was the skeleton of a man.

Forensic anthropologist Walter H. Birkby examined the skeleton in Tucson and found a .32-caliber bullet lodged inside the cranium and an entry hole near the right ear. Likely the man had been murdered. Less likely, he had committed suicide. What the hell was Emery Kolb doing with this skeleton in his garage?

The discovery of this skeleton mushroomed into a classic case of people seeing what they wished to see. It quickly raised the suspicion that collusion existed between Bessie Hyde and Emery Kolb. Did he rescue her from a maniac he believed would drown her otherwise? Clearly Bessie Hyde was afraid to continue down the Colorado. She knew that the worst dangers lay below

Diamond and that Diamond was her last hope of escape, because Glen refused to abandon the trip. Was Liz's story about stabbing Glen a cover-up to protect Emery Kolb? During Liz's 1971 trip, Emery was still living at his studio on the South Rim. Was Scott Thybony correct in observing that, by taking this trip, Liz, as so many murderers have done, was being drawn back inexorably to the scene of the crime? Did Emery Kolb play any role in the disappearance of the honeymooning Hydes?

Walter Birkby heard the rumors and found it "preposterous that he [Kolb] would off somebody and then keep the damn body around." Birkby's anthropometric measurements found the skeleton too stocky and five years too young to be that of Hyde. He reconstructed the face. The chin was too square, the cheekbones too wide, and the eye orbits angled wrong for Hyde. Birkby concluded: "These bones that were submitted for analysis are not the remains of Glen Hyde. Period."

Emery Kolb's housekeeper explained that Emery had found the skeleton at an old mining site. Art Gellenson noted that Kolb had had the skeleton so long that he had even donated it to the local school for their biology lab but had eventually received it back again. Norm Tessman, the museum curator in Prescott who analyzed the clothes found with the skeleton, pronounced them vintage twenties or thirties, but the belt buckle was not the one Glen had been wearing in photos of him from the trip. So Glen Hyde was still missing and Emery Kolb was still a hero.

Despite all this, the skeleton took on a life of its own, drawing in Hyde's family and several other people, including Arizona's Canyon-phile governor of the time, Bruce Babbitt. But during a Canyon trip we ran just prior to the 1988 presidential election, Babbitt admitted to me that he had learned almost nothing from the investigation he ordered.

Marty Anderson, leaving no stone unturned, finally contacted Earl Helmick, Bessie's first husband, to tell him the television series "Unsolved Mysteries" was airing a program on Glen and Bessie Hyde. Over the years Helmick had refused to talk about Bessie to anyone. Now, in his eighties, he told Anderson too, "I don't want to talk about it." But then he explained why: "Bessie is dead."

Is she? What actually happened to Glen and Bessie Hyde? It beats me, but I would love to write the screenplay for the movie.

On board Glen Hyde's scow, along with the rest of their gear, the Kolbs and Brooks found Bessie Hyde's purse. They gave it to the elder Hyde. It still had a few dollars in it. Would any woman voluntarily hike out of Grand Canyon, into a world that did not know her, without her purse? On the other hand, could any passenger on the Colorado River claim to be Bessie Hyde, then tell four professional guides in detail how she had murdered her husband with a knife forty-three years earlier, then forget the whole story?

14
Radioactive Oasis

In the desert, water is everything, life itself. Nothing is more precious, more revered, more contested. To survive here on a hot day, a human must gulp two gallons of the stuff. Being stranded here without water leaves one roughly the same chance of survival as being marooned on the moon without a space suit.

This reality makes Havasu Canyon an even more beautiful oasis than if it were located somewhere in the green lands east of the hundredth meridian. But no matter where Havasu might be located, it would be a Shangri-la. When I first hiked from Mile 157 at the bottom of Grand Canyon upstream along the aquamarine creek flowing and tumbling between the thousand-foot walls of pink Muav, Temple Butte, and Redwall limestones of Havasu Canyon, I was stunned by its beauty. My brain seemed reluctant to file the images I was seeing as genuine. "Are you hallucinating?" it seemed to accuse. Each bend of the wide, U-shaped canyon revealed a new roaring waterfall spilling blue-green water over curving red terraces of rimstone dams flanked by lush wild grapes and flickering velvet ash. No place on earth could be this beautiful and be real.

Perhaps my skeptical neurons were right. As a test I dropped my pack and knife belt and leapt into a pool. Blue-green water thundered and bubbled around me, cool and embracing, a rescue from the hundred-degree temperature I had just escaped. I tumbled deep in the gentle flow for a moment and surrendered all responsibility to Havasu Creek. The water carried me toward the Colorado. Bubbles raced skyward in vanishing curtains to reveal algae growing everywhere on the travertine formations forming within the creek. I drifted underwater toward the next rimstone dam formed of travertine deposited—one molecule of calcium carbonate at a time—on tree limbs once lodged in the creek. The wood was now gone, but the travertine preserved casts of those limbs faithfully within its dams. I pushed off the gravel bottom and shot to the surface. I spun in a backward somersault to stop with my feet against the travertine dam. Then, as Havasu Creek rushed past me to thunder downward into yet another pool, I stared at the scene. It was too big and strange to fit into any cerebral category, but more atavistic neurons told me instantly where I was. Paradise.

Paradise here equals water. Once upon a time, perhaps fifteen to forty million years ago, this canyon held a large river that flowed from mountains far to the south near present-day Prescott. In those days before the uplift of the Colorado Plateau, the area to the south was higher in elevation, and this was lower. Present-day Cataract Creek (aka Havasu Creek) was then a river draining thousands of square miles of those southern mountains. This explains why "oversized" Havasu Canyon is fully as wide, and in many stretches even wider, at creek level than today's Grand Canyon. Today the topography of Northern Arizona has reversed, and the landscape here slopes away to the south. So do the streams. And instead of being a river, Cataract Creek drains only the seasonal flash floods running north off the nearer portions of the Coconino Plateau. While such flash floods occasionally rival that long-dead river, almost all of what flows down Cataract-Havasu Creek is artesian water from Havasu Springs in the Supai Formation. It is this blue-green water that sustains the Havasupai.

It has also drawn hordes of ecological pilgrims to Grand Canyon's Shangri-la. One early pilgrim hiked down from Topocoba Hilltop fourteen miles to Supai Village in the early 1950s during the monsoon season. He continued another half dozen miles downstream past Navajo Falls, Havasu Falls, then Mooney Falls (named after a prospector named Daniel W. Mooney, who in 1880 fell to his death from the end of his rope while attempting to descend beyond the two-hundred-foot falls to search for valuable minerals in Lower Havasu). Dazzled by the beauty and immensity of the falls and hypnotized by the primeval wildness of the canyon and its unearthly little river, the ecological pilgrim stayed for a month. What did he find to do for thirty-five days? "I agonized over the girls I had known and over those I hoped were yet to come. I slipped by degrees into lunacy, me and the moon, and lost to a certain extent the power to distinguish what was and was not myself."

But these existential conundrums were the easy part. His urge for exploration became a predator that turned on him. He hiked up a large tributary canyon of Havasu, Beaver Canyon I suspect, to the Esplanade bench atop Supai. There he realized that he had too little time to retrace his several-mile route back to camp at Mooney Falls before nightfall. As the raven flies, he was less than half that distance away. So he sought a shortcut back into Havasu. Had he known that routes into Grand Canyon through the Redwall cliff below the Supai, and even through the Supai itself, are as rare as good politicians, he never would have been seduced by his chimerical shortcut. But he did not know. He was new here. So he descended, carrying neither rope nor water, into a steep, shadowy, extremely narrow defile in the Supai.

Eventually he reached the lip of a twelve-foot drop into a pool cupped by vertical walls in the sandstone. The decision was agony. If he dropped into this pool but found the section below impossible to descend, he would be trapped, because climbing back up from the pool also appeared impossible. But he could not retrace his original route before morning—which dictated spending a very unpleasant night here on cold, hard rock. Worse, if it rained and he was stuck here, he would become part of the flash flood. Certain death.

He inched over the rim and dropped feet first into the pool. Then he dropped into a second pool and peered over its edge. Beyond it was an eighty-foot drop onto a rock pile. Certain death.

The long view down to the pile of boulders spurred utter panic, tears, self-recriminations, crazy plans to shred his old jeans and T-shirt and weave them into a rope, then crazier plans to use them instead to light a signal fire that no one would see. Finally, after contemplating the infinite stupidity of his fate, the few loose stones scattered in his sepulcher gave him another idea. He took each, one by one, and swam with it back to the slickrock chute standing between him and the rest of his life. There he made a shaky pile two feet high, climbed it, strained with all his might to find some finger- or toehold. But he failed, crept back down, and started to cry again.

New ideas were now as rare as IRS errors in your favor. But through his tears he noticed his walking stick lying nearby. He swam back and propped it on his pile, placed his bare big toe on the top of the stick, then painfully inched up the polished Supai sandstone in a ludicrous balancing act impossible without death whispering in his ear. The stick allowed him to reach higher up the S-shaped chute where friction between the rock and his wet clothing helped him to inch ever so slowly higher. Finally, the moment of truth. A last painful lunge with his big toe against the walking stick sent the stick clattering into oblivion and the stones collapsing into a state of increased entropy. But he "crawled rather like a snail or slug, oozing slime, up over the rounded summit of the slide."

His triumph was short-lived. Ahead was an overhanging spout twelve feet above a deep plunge pool that he knew was impossible.

Here there was no stick, no rocks, and no magic tricks. Swimming underneath the drop-off to flounder and scrabble at the slippery rock eventually convinced him that a marine assault *"was not the way."* He swam back to solid ground to lay down and die in comfort.

Instead he studied the walls nearby. Every weathered chink, crack, bump, curve, pock, and ledge. He saw that the walls formed almost a square corner where they joined the face of the overhang. The minute crevices and inch-wide shelves on either side hinted that escape might be possible.

After regaining some nerve and steadiness, he got up off his back and tried the wall, inching like a crab with fingers and toes to traverse above the pool. His soggy boots, dangling from his neck and swinging with every movement, threw him off balance and sent him plunging into the pool again. He swam to the edge, disgusted. There he unslung his boots and tossed them up over the obstacle ahead. Dead men don't need boots, but if he lived, they'd be there:

> Once more I attached myself to the wall, tenderly, sensitively, like a limpet, and very slowly, very cautiously, worked my way into the corner. Here I was able to climb upward, a few centimeters at a time, by bracing myself against the opposite sides and finding sufficient niches for fingers and toes. As I neared the top and the overhang became noticeable I prepared for a slip, planning to push myself away from the rock so as to fall into the center of the pool where the water was deepest. But it wasn't necessary. Somehow, with a skill and tenacity I could never have found in myself under ordinary circum-stances, I managed to creep straight up that gloomy cliff and over the brink of the drop-off and into the flower of safety. My boots were floating under the surface of the little puddle above. As I poured the stinking water out of them and pulled them on and laced them up I discovered myself bawling again for the third time in three hours, the hot delicious tears of victory. And up above the clouds replied—thunder.

Young Ed Abbey crawled back to the upper Supai and hid under a tiny ledge to sit out the all-night storm. Behind him, in that steep, shadowy, extremely narrow defile in the Supai, water roared and plunged through the dark like a storm drain from hell. It was one of the happiest nights of his life.

One day twenty-two years later, I hiked twenty miles down from Hualapai Hilltop to the Colorado. My son Cliff was five days old. I had left Connie and him in Flagstaff in the hands of Marian Spotsworth, a close friend who also was a nurse. I was hiking to relieve Stan Boor, who had rowed my boat sixty-nine miles down from Phantom Ranch, where I had hiked out to be with Connie when Cliff was born. We passed the baton at Mooney Falls, where I thanked him for making the impossible possible. That night we camped downstream from Havasu a few miles, where I cooked dinner for the group. It was my night to cook, but cooking was tougher than usual because a few days earlier one of our two stoves had caught fire, and Fabry had tossed it hissing into the river. Anyway, my point here is that hiking Havasu from rim to river these days is something many of us take for granted. Now it looks as if none of us should.

Havasu is the most beautiful of the many beautiful creeks I have seen in Arizona, or anywhere in the Southwest . . . or anywhere in the world, for that matter. It radiates an aura of purity, of innocence of the chemical abuses suffered by most other streams in the United States. It seems a holy place hidden purposefully from the follies of the industrial revolution, perhaps as a seed crystal to replenish the beauty of our planet once we have ruined the rest of it. If so, the gods are in for a shock.

Now the Havasupai worry that Havasu Creek will soon become radioactive. No, our boys with the Manhattan Project did not sequester an ultra-top-secret lab down here among the Havasupai to conduct clandestine experiments. At least I don't think they did. But this part of the Colorado Plateau is rich in uranium. Too rich. Entirely too rich for the welfare of the mysterious Havasupai.

Back in the 1950s Douglas Schwartz made his way down into

the Supai Valley of Cataract Creek in Havasu Canyon to spend six months studying the Havasupai. He had a lot of questions, like where the Havasupai had come from. He knew that around A.D. 600, before the rise of the Anasazi culture, people showed up on the Coconino Plateau. Archaeologists call these people Cohonina (a Hopi term meaning Those Who Live in the West). For a couple of centuries the Cohonina apparently eked out a living by hunting and gathering on the plateau, and also by planting the triad of southwestern native American crops—maize, beans, and squash—for added security. By A.D. 900, contemporary with the Pueblo II Anasazi, a few Cohonina had taken up residence in Havasu Canyon below its confluence with Cataract (Havasu) Creek in Supai Valley ten miles upstream from the Colorado. Besides irrigating, they hunted and gathered in Havasu down to but not beyond the impassable Mooney Falls. Two centuries later, a sudden change occurred: the people seemed to abandon the plateau while at the same time the population of Supai Valley exploded by several hundred percent. Why?

As with the Anasazi to the east, the cooling climate may have forced the Cohonina from the too-cold plateaus. But because at this same time, A.D. 1100, people began building cliff dwellings away from the valley floor (which was subject to periodic flash flooding) so intensively that "every available ledge became a home," Schwartz suspects that something new was happening. What?

Raiders. Schwartz suggests the following as evidence that another tribe began preying on the people we now know as the Yuman-speaking Havasupai (People of the Blue-Green Waters): they abandoned their plateau homes, their population exploded in the remote and isolated valley, and they suddenly began building defensible cliff dwellings with an unprecedented vengeance. Two centuries later these cliff dwellings were abandoned. It is unclear whether these early cliff dwellers were ancestral Havasupai or instead were unrelated Cohonina with Anasazi affiliations. But by 1300 no more cliff dwellings were being used. From

216

that time onward, people who almost certainly were ancestral Havasupai built thatched houses on log and beam frames on the valley floor, where the population had plummeted to its size in 1100. These Havasupai apparently lived a dual life: farming during the summer in their valley, then hunting deer, elk, pronghorn, and desert bighorn and gathering piñon nuts on the plateau in winter.

For centuries the Havasupai lived like this while maintaining excellent social and trading relations with the Hopi to the east and with the closely related Hualapai just to the west. They were in a constant state of friction, however, with the Yavapai to the south and with the Apache, and intermittently with the Navajo.

The first white man known to have visited the Havasupai in their canyon was Fray Francisco Garcés, guided by Hualapais in July 1776. Having no interpreter, the solitary Franciscan failed to gather the Havasupai into the protective fold of the church, but he noted that at that time the Havasupai already had horses, cattle, peach trees, cloth, and other European goods traded from the Hopi. Five years later Garcés was killed by Yumans in the south.

Not surprisingly, when the Anglo settlers arrived in northern Arizona, they quickly claimed the best lands of the Havasupai. By 1882 the whites had legally shoved them all back into Havasu Canyon on a 518-acre reservation. This is less than one square mile of Supai Valley—a considerable decline from the 6,750 square miles of territory they had occupied when whites first arrived. A few Havasupai remained at Indian Gardens, along Bright Angel Trail from South Rim Village to Phantom Ranch. But, as W. E. Garrett reports being told by some Indians, "President Roosevelt personally asked them to leave [in 1911] because 'he was going to make it a park for everyone.'"

Considering the ninety years of prospecting by whites in Havasu Canyon after 1866, it may be lucky that the Havasupai were not bilked of those last 518 acres as well. In the early 1970s they fought back in court with the help of anthropologist Bob Euler. In 1975 they won by congressional award an expansion of their reservation to 160,000 acres (250 square miles), 95,000 acres (148 square miles) of which is in Grand Canyon National Park but is available for their use.

Today approximately 500 Havasupai live permanently or intermittently in their beautiful oasis. Their primary income these days comes from tourism: fees and so on from people hiking, horseback riding, or helicoptering into their reservation. Because Havasu Canyon is the most dazzling paradise in Arizona, tourism will yield a steady income, but although they are interested in money and the things it can buy, most Havasupai seem to want little to do with the spiritual realm of whites. Who can blame them?

One might guess a couple of things about their life in Havasu Canyon. The first would be that life is idyllic; the second, that it is boring. Not much happens beyond sporadic influxes of tourists hauling in their strange notions and mountains of possessions. Within the tribe there are problems of alcoholism and marijuana abuse, plus occasional adultery and rare violence. What do these people look forward to? What do they worry about?

Recently they started worrying about subatomic particles. Havasupai worry that their water supply, the blue-green spring water flowing down Havasu Creek, will become polluted with radioactive uranium and will no longer support normal life. They also worry that Red Butte, the region they hold most sacred—it is as significant to them as the Sipapu along the Little Colorado is to the Hopi—will be damaged beyond the powers of the universe to heal.

Why? For many of the details of this issue I owe a debt to Flagstaff environmental journalist Dan Dagget and to Kathleen Stanton. The source of the radioactivity problem is geological: there are thousands of breccia pipes scattered on the Colorado Plateau on both sides of the Grand Canyon. Breccia pipes here are caused by the collapse of overlying strata—say of Supai, Hermit, and the others all the way up to the Kaibab and beyond for an additional vertical mile of Mesozoic formations that have now

218

vanished due to erosion—into giant solution caverns that have formed beneath them in the Redwall aquifer. These breccia pipes tend to store collapsing rock, funnel groundwater, and concentrate many minerals that are far more dispersed in the surrounding bedrock—minerals that were present in those now-vanished upper strata.

According to George Billingsley, at the end of World War II the Hacks Mining Company worked an old copper mine in Hacks Canyon, a tributary of Kanab Creek, but abandoned it because the road washed out. In the 1950s someone identified torbenite, a uranium-copper mineral, in the ore from the Hacks Mine. In the late 1950s, the Rare Metals Corporation of America worked the same mine, probing two adjacent breccia pipes, then they too eventually abandoned it. In 1980, Dagget reports, Energy Fuels Nuclear (a Denver-based company backed by two Swiss utility companies) found an apparently new breccia pipe in Hacks Canyon. Their head engineer described it as the richest deposit of uranium in the United States.

Additional prospecting, Stanton adds, revealed several breccia pipes scattered over the plateau, containing sedimentary uranium concentrated during the millions of years when sheet erosion was stripping away the overlying uranium-rich strata of the Chinle Shale. These uranium glory holes, which may extend thousands of feet down to the base of the Redwall, contain ore three to ten times richer in uranium than any other ore known in the United States. Energy Fuels Nuclear refines it to uranium oxide, U-308, called yellow cake. Even though the price of yellow cake dropped from forty dollars a pound in 1978 to seventeen dollars a pound in 1988, the concentrations of ore in these breccia pipes make mining them worthwhile. In fact, Stanton reports that in 1987 alone Energy Fuels Nuclear, which has located nine uranium-rich breccia mines on the plateau, grossed $85 million from 5 million pounds of yellow cake taken from just two of their mines along Kanab Creek. This was more than half the total produced in the entire country.

Dagget notes in warning that six of Energy Fuels Nuclear's new mines have been opened from five to sixteen miles from the boundaries of Grand Canyon National Park. This company's aim apparently extends beyond merely making money by selling uranium to monopolizing the richest uranium resources in North America. Of the million-plus acres of public lands on the plateau opened to mineral exploration by the archaic 1872 Federal Mining Law (which allows—at a cost of a mere $2.50 to $5.00 per acre—mineral claims and which is still in effect despite the fact that 670,000 acres of this, our land, is a roadless area under consideration for designation as a federal wilderness area), nearly 700,000 acres have been grabbed by Energy Fuels Nuclear's 40,000 mining claims. They anticipate operating fifty mines in the next thirty to forty years in the Grand Canyon area. Their claims comprise an area four and a half times greater than the Havasupai's newly enlarged reservation (perhaps the Havasupai should apply for Swiss citizenship). The rest of the land has been claimed by Johnny-come-lately mining companies.

Where does all this uranium go? Into bombs, one wonders? Well, one must continue to wonder. According to Stanton, Energy Fuels Nuclear spokesperson Pam Hill equivocates, admitting that 30 to 40 percent of the yellow cake they sell goes overseas but refusing to divulge to what nations and for what purposes. That information remains a company secret. This is an interesting twist of liberty in the Land of the Free. By contrast, as part of uranium-rich Australia's stand against the proliferation of nuclear weapons worldwide, it refuses to export uranium unless its destiny is known. If it is for other than peaceful uses, Australia disallows export.

Be that as it may, the Havasupai believe they face two major threats from Energy Fuels Nuclear. The first is the threat of radioactive contamination from the company's three new mines proposed south of the Colorado River on the Coconino Plateau and in the drainage of Cataract Creek: the Platinum, the Scorpion, and the Canyon mines. Because of their more critical location in the drainage of Cataract Creek, the first two were denied by Bob Lane, Arizona State Lands Commissioner during Bruce Babbitt's administration. After Evan Mecham gained office, Energy Fuels Nuclear refiled the proposals. But their refiling effort has fizzled out; the mines are simply too obvious a threat to the Havasupai's water supply, among other things. The third, the Canyon Mine, though hotly contested by environmentalists and the Havasupai, is being developed as planned. Do such mines pose a real threat to the rarest of all resources in this desert—clean water?

Ask the Navajo. Several uranium mines have already caused health problems on their reservation. Stanton notes that in 1979 the tailings dam of the United Nuclear Mine in Churchrock, New Mexico, failed and dumped 94 million gallons of "radioactive slop" that flowed into the Little Colorado and on into Grand Canyon. Ah, one might argue, but that was a catastrophic quirk unlikely to happen again. But routine mining in the Grants Mineral Belt of New Mexico, in which the uranium is contained within the aquifer strata, required regular pumping to keep the mines workable. Twenty years of this wet mining, with no accidents, dumped more radioactive material into the Rio Puerco and into Navajo livestock—and into Navajos—than did the United Nuclear Mines accident.

Radioactive tailings are not an ephemeral threat. As uranium decays into thorium, radium, polonium, and lead it gives off a deadly stream of alpha and beta particles and gamma radiation. The final decay to nonradioactive lead takes about 40 billion years. Nor is this radioactivity a theoretical threat. Physiologist Al Goodman found that the rate of birth defects among Navajos in the nearby Cameron area along the Little Colorado north of Flagstaff, where uranium was mined thirty years ago, is five times higher than the national average. Dr. Donald Calloway found ovarian and testicular cancers in Navajo children around Shiprock to be fifteen times higher than the national average. Many other cancers and birth defects are also much higher among Navajos living in uranium mining districts than Navajos elsewhere.

The mines proposed and operated by Energy Fuels Nuclear on the Coconino Plateau are dry mines that involve excavating those dry "hot" breccia pipes down to the Redwall Limestone, the predominant aquifer stratum here. But, unfortunately, Energy Fuels Nuclear's operations are not immune to problems. James Mahoney and Phyllis Hogan report that at the company's North Kanab Uranium Mine, "a minor flash flood recently tore down their precautionary berms and sent the deadly gruel flowing down through the Kanab Creek Wilderness to the Colorado River." The "gruel" was four tons of high-grade uranium ore. No one bothered to measure radioactivity levels to determine how far that washout traveled. So the question boils down to this: How serious a threat is the Canyon Mine?

Supervisor Leonard Linquist of the U.S. Forest Service (Williams, Ariz.), who holds jurisdiction over the Canyon Mine site near Red Butte and Tusayan, filed an environmental impact statement. Yes, the mine would affect use of the area by elk, possibly affecting their migration or calving. And yes, some dust raised by the ore trucks, carrying a projected 600,000 pounds of radioactive ore annually, might cause slight air quality problems in Grand Canyon National Park (whose boundary is only seven miles away). But the trucks themselves allegedly present little hazard—despite an Energy Fuels Nuclear ore truck spill in 1986 on the Navajo Reservation reported by Mahoney and Hogan, who also noted that the company's personnel were caught attempting to "clean up" spilled ore by covering it on-site with sand. But it would require, Linquist's environmental impact statement points out, a monumental flood near the mine to wash radioactive tailings into the drainage of Cataract Creek. In short, "no significant environmental impacts are expected from mining operations and ore transportation."

Shockingly, Energy Fuels Nuclear's Groundwater Quality Protection Permit from Arizona's Department of Environmental Quality even allows the miners to self-monitor the presence of radionucleotides and hazardous materials leaking into the environment.

But beyond whatever threat the Canyon Mine holds for contaminating the water in Havasu Canyon, the Canyon Mine is a thing of evil to the Havasupai due to its proximity to the unique, imposing island of Triassic rock called Red Butte. Strangely, this was apparent to no one outside the tribe. When the Havasupai claimed that this was true, Energy Fuels Nuclear reacted with disbelief. The anthropologists, they said, knew that the Havasupai were spiritually impoverished. Douglas Schwartz's 1983 analysis of Havasupai religion, for instance, concurs with an earlier anthropologist, Leslie Spier, in saying that "religion and ceremonialism were not highly developed and occupied only a minor place in Havasupai life." So it is not surprising that Energy Fuels Nuclear reacted to the Havasupai protest over Red Butte by asking, what religion?

Kathleen Stanton reveals the missing anthropological data:

> Last February [1987], five members of the Havasupai
> Tribe made the long journey to Washington, D.C., and there
> met with Forest Service officials. Their message, one they had
> kept hidden from white eyes until now, is that the mine
> Energy Fuels proposes to sink thirteen miles south of Grand
> Canyon National Park [from the entrance] is located directly
> on the most sacred ground in their universe [Red Butte].
>
> To sink a mine shaft at the Canyon Mine site is to rip a
> hole in the abdomen of their spirit mother, who each year
> gives birth to renewed life and places it briefly on her abdomen
> to rest before it begins its journey, the Havasupais explained.
> The site has separate significance to the Hopi, who believe
> it lies in the path of a spirit bearing good will to the Havasupai,
> their trading partners since ancient times.

"The center of our belief is there, right there," said Clark Jack, one of the four Havasupai elders at the meeting. "Everything starts from there and will return from there, such as the [child when it is grown], because he always turns back to the Mother."

The Havasupais said they did not speak out earlier because to speak out at all violates a basic tenet of their religion. . . . "No one is going to go out and say, 'This is my personal religion,' because it is something we grow up with," explained Rex Tilousi, another Havasupai elder who made the Washington trip. "It is something we have to attain, it is something that we have to mature into."

Protests against the Canyon Mine by environmentalists in Grand Canyon National Park in 1988 resulted in the arrest of twenty-nine people for civil disobedience. Dagget explains that the protesters were "asking for an area-wide study of the possible effects of the mining developments and a moratorium on all mining in the area until that study is complete. They are also asking for a congressional investigation into alleged collusion between government land managers and mining company officials, and a repeal of the 1872 Mining Law which they say gives mining companies a carte blanche to take possession of public lands." Stewart Udall sums up the inadequacies of the 1872 mining law: "On the public estate mining is still a search and destroy mission."

The Havasupai Tribe is pressing a suit against the U.S. Forest Service on the grounds that it is allowing the desecration of a sacred site. But Energy Fuels Nuclear has already invested millions in the development of the Canyon Mine, and as I write this, work on the mine is going full steam despite appeals filed by the Sierra Club, the Arizona Wildlife Federation, Canyon Under Siege (a Flagstaff environmental group), Friends of the River, the Tonantzin Land Institute, and the Havasupai. If the appeals are unsuccessful, environmentalists will take the case to federal court to challenge the Forest Service's environmental impact statement that found no significant impact.

Victoria Harker points to an ominous note in the Havasupai's suit over the Canyon Mine on the basis that "Red Butte, four miles south of the uranium mining site, is the navel of the Earth, where the Earth was connected with the Universe. . . . located in the abdomen of the Earth where the 'life spirit of renewal' is attached." On April 20, 1988, she notes, the U.S. Supreme Court overturned by a vote of 5 to 3 a ruling by the Ninth Circuit Court of Appeals (in San Francisco) forbidding timber harvesting and road grading in Six Rivers National Forest in California, which is considered sacred by the local Yurok, Karok, and Tolowa tribes. The Supreme Court defended its decision to open Six Rivers to logging by pointing out that nothing in the U.S. Constitution justifies allowing Indians ultimate rights over public land. The issue of how that land became public land was not considered in their decision.

I shipped my oars for a moment. From the Colorado River the confluence of Havasu is obvious if you know the geography, but pulling out of the mainstream into the narrow mouth of Havasu Canyon still requires stopping on a dime. Neophytes often see it too late to stop. Now I rowed so close to the polished, fluted, wall of dark gray limestone that our stern seemed in danger of being sliced to ribbons. I pulled on my upstream oar just before reaching the mouth to tug my boat into the small eddy . . . on that dime. Behind me, spaced at long intervals, four boats followed. The strangely blue-green waters of the little eddy here were empty; Havasu was ours.

After rigging the boats to the cliff face to withstand a flash flood disgorging from the narrow slice of bedrock funneling Havasu into the Colorado, we ascended ledges of Muav and followed a narrow trail into the mouth. We passed a solitary ocotillo (Spanish for "little torch"), a spiny octopoid of a plant waving crimson-tipped blossoms ten feet in the air as if announcing, "Welcome to the Lower Sonoran Life Zone."

The immense, U-shaped canyon was still in shade. I led

everyone across the turquoise creek where it is less than waist deep. Later we followed a path worn a foot deep into the hard-packed sand and flanked by grass so lush that every step is caressed by greenery. Ahead two western whiptail lizards rushed before me frantically. Tiny dinosaurs fleeing in terror from a voracious *Tyrannosaurus rex* . . . me. The forest of grass along the path is so impenetrable that the frantic lizards jetting through the dust found no exit route to escape me, so they fled along the path, my footfalls even with their pace and hot on their tails, which were dragging lines in the fine sand. One hesitated. I slowed down to avoid stepping on it. Finally they located an exit and dived from the trail to rustle through dry leaves. As soon as they vanished into safety, they were replaced by new whiptails who began fleeing ahead of me in a panicked search for an exit off the trail. And so on.

The morning was so peaceful that no one uttered a word. As if on a holy pilgrimage, we trekked purposefully upstream past deep, inviting pools created by the travertine cementing of log jams, along a trail winding through acre after acre of native wild grapes, tall grass, scattered groves of mesquite and catclaw, and through thick stands of velvet ash, willows, and old cottonwoods gnawed by beavers. I scanned for bighorn sheep and spotted several prince's plumes, concentrations of which indicate uranium in the soil.

Our destination was Beaver Falls, four miles from the Colorado. Beaver is a paradise within a paradise. Tandem twenty-foot falls feed one wide, deep pool, then another and another; some are fifty yards long and twenty feet deep. These swimming holes exceed all specifications for Paradise. At the base of the lower falls is an optional excursion into the limits of one's courage—or faith. It is called the Green Room. And I definitely do not recommend it.

The Green Room is under a split between the two main streamers of the lower falls, perhaps six to eight feet directly upstream behind the wall of travertine behind the falls. To enter it one must jump into the boiling pool between the streamers, swim to the wall, and, as water pounds on one's head, locate the base of the apron of travertine about four feet underwater, then inhale a very large breath of air and dive. That's the easy part. Next, after diving about four feet straight down, one must swim underwater directly under the falls and the rock, away from the atmosphere, the sunlight, and everything that makes life possible, for about eight feet. To surface too soon during this swim is to collide head first with the sharp ceiling of travertine, airless and dark and suddenly evil, an easy place in which to drown. One must swim those eight feet horizontally and then surface with one hand placed on one's head to prevent lacerations by stalactites. When that hand breaks into air, it's time to surface. Carefully. If the sun is shining on the pool outside, this tiny cave hidden beneath the falls will glow a soft, radiant green.

On my last hike to Beaver Falls I was leading twenty main players in the Hollywood movie industry. It was an afternoon hike, and the sun had already fled from the pool and falls leading to the Green Room. Only one person was interested in exploring it, Hollywood superstar Tom Cruise.

I explained to Tom exactly how to get in. Then I leapt into the pool between the pounding streamers to lead the way. He plunged in behind me but then gave me fifteen seconds lead time. I swam through the murky water, not quite jet black but too dark to see even my own hands. You've got to be nuts to do this in the dark, I thought, or you must have faith in your comrades. After a year or so my hand finally felt air. I surfaced, then turned quickly to watch for Tom. On anyone's first excursion in here it is impossible to know when to surface. I would guide his head with my hand.

But, with sunlight gone from the pool, the light was too dim. Was that him? It suddenly struck me: Here was a guy with a face worth at least one hundred million dollars swimming face first in the dark into a black room where everything but me was sharp. Tom Cruise was currently the hottest, most valuable property in Hollywood, and despite this he was a good man: generous, humble, with a great sense of humor, a positive attitude, and the mental and physical self-discipline to accomplish his goals. I liked him. But I had just sold him on a risky experience that might rearrange his face.

I reached deep into the water. That had to be his head. I grabbed his hair as if he were just about to be swept over Niagara Falls and tugged him into the safe zone. Clean, not a rock grazed in passing. Relief.

Ignoring my iron-gripped panic, he grinned that hundred-million-dollar grin and said, "This is *great*, Doc!" as he studied the not-so-Green Room. We were surrounded by stalactites clustered like the teeth of Monstro the Whale swallowing Pinocchio. Our pool was apparently bottomless, but we barely had head room. "You can't even hear the falls," Tom whispered, almost in awe. Suddenly I respected him even more. And I knew that I need not worry about his roaring exit through the falls to the outside world. He had what it took.

While the spiritual significance of Red Butte near Tusayan came as a revelation to whites after so many years of assuming that the religion of the Havasupai consisted almost entirely of simple shamanism, their religious life is far more perplexing than I have so far hinted.

In 1974, Haile Selassie, the last hereditary emperor of the ancient Coptic Christian empire of Ethiopia, King of Kings, Elect of God, and Conquering Lion of the Tribe of Judah, was either allowed to die of natural causes or died due to poison in his food while imprisoned under house arrest in his palace in Addis Ababa by the revolutionary Marxist forces of the Dergue. At eighty-one years of age, Haile Selassie, better remembered by some by his precoronation name, Ras (Prince) Tafari, vanished corporeally but left behind a spiritual legacy. As Mengistu Haile-Mariam, leader of the revolution, eliminated other members of the royal family and numerous independent-thinking members of his own revolution, a small group of Ethiopians canonized Ras Tafari based on a Coptic biblical prophesy written in Amharic identifying him as the son of God. These people became known as Rastafarians.

Reggae music associated with Rastafarian refugees in Jamaica now commonly drifts from the windows of the prefab housing on the Havasupai Reservation. Reggae and its message have created a rare ripple of response among many of the younger Havasupai—who live in a society where unemployment, alcoholism, and a lack of opportunity are serious problems. Reggae also has an irresistible beat. Part of reggae's Rastafarian message aims at social harmony, equality, ethnic integrity, and antimaterialism. Much of the rest of it is antiwhite and predicts the fall of Western civilization and of the "Babylon" of North American white culture. For many, reggae is religion. Marijuana is its sacrament.

How did Rastafarians connect with these Southwest Indians? The thread was thin. One story is that in the 1970s three California Indians appeared at Hualapai Hilltop at the trailhead to Supai and asked some local Havasupai, "You ever heard reggae?" Tobey Watahomigie said no and bought one of their tapes, Bob Marley's *Rastaman Vibration*. This tape was the seed crystal for a new musical, spiritual, and religious movement eight miles down the trail in Supai Village.

In 1982, Chris Blackwell of Island Records, manager of reggae giant Bob Marley, visited a record store in Las Vegas and saw a cluster of Native Americans buying fistsful of Marley tapes. Curious, Blackwell asked who they were. Indians who lived in the bottom of the Grand Canyon, one told him. He was Bob Marley's manager, he informed them. They loved Bob Marley, the Havasupai said. Blackwell told Bob Marley's mother, Cedella Booker. The wheels turned. The Havasupai Tribal Council invited Booker to Supai Village. She traveled to Supai—Little Jamaica—where she was treated as royalty. She danced to her son's music in the thunder and mist below Havasu Falls.

Some Havasupais apparently consider Bob Marley almost a god. Photographs of him are everywhere—from bedroom walls to T-shirts to the radiator of the only tractor in Supai. Most young Havasupais are addicted to reggae. It was probably inevitable that a reggae band would find its way into the Shangri-la of the Southwest. In 1983, again at the invitation of the tribal council, the band Shagnatty hiked in. A helicopter carried the equipment. Beth Trepper, also cleared by the tribal council, went as a photojournalist to record one of the strangest spiritual events in southwestern history.

Everything went fine on the night of the concert, Beth told me. Havasupais even leapt onto the stage to join in making music and to toast the band. But in the morning, Trepper explained, when they were about to helicopter back to "Babylon" for Shagnatty's next gig, she decided to catch the last flight so she could take more photographs. It was a mistake.

Chief Clark was drunk. He dragged Beth into the cafe, where he yanked her camera from her and threw it on the floor, telling her, "You're a black journalist, and you're going to write bad things about us." He confiscated her film and added, "And you cannot leave."

At first she thought he was joking, but no one was smiling. When she looked out the window at the path leading out of Supai, it was blocked by Havasupais on horseback. If this was a joke, it was pretty elaborate. Clark was not going to let her leave Supai.

Trepper tried to explain that she would only write good things (which she did—this story she later told me personally) and that he was mistaken about her. The drummer and manager of Shagnatty came in to tell her the chopper was waiting. Beth appealed to them to intercede for her. They said they had to leave right away to make it to their next gig. If they waited much longer, the band would miss it.

230

Finally the pilot appeared. He dispelled her last hope that any of this was a joke. Clark was a serious man, the pilot told her, and none of this should be taken lightly. Finally Clark demanded money to let her leave. The grand total of cash that the four could scrape together was $39. But it was enough. Clark was appeased. The pilot eased Beth out, even got her film back from Clark, and belted her into his machine.

Below Trepper the little oasis and thousand-year home of the Havasupai shrank and then was hidden by the cliffs of Supai Sandstone. Somewhere down there Bob Marley's voice drifted from the window of a prefab house. She had trouble connecting her violent experience with the paradise below.

Two days later two Havasupais were mysteriously murdered.

Years later the Havasupai people impeached Chief Clark Jack—as I write this he lives in exile, wandering the streets of Flagstaff.

The recent meteoric rise of the Rastafarians and reggae fans among the Havasupai is not an isolated case. It has spread to the Hualapai, the Hopi, and even to the Navajo, though less strongly. Some of the older Havasupai now fear that it is already too late to prevent a spiritual schism in their tribe. This schism divides generally the older traditionalists and the young Rastafarians. If this happens, much of the traditional relationship between the Havasupai and the rest of the universe, which has held firm for the past thousand years, may be lost. But perhaps Rastafarianism will be no more durable than a flash flood roaring down Havasu—like the fifteen-foot-high wall of water that during September 1990 displaced a hundred Havasupais and scoured 90 percent of the riparian trees out of Havasu Canyon—a phenomenon that today has force and influence but that is forgotten as the years pass.

Should we worry about this most captivating paradise in the Southwest, this unique oasis where water is everything? Both the threat of radioactive pollution and the spiritual turmoil only a half dozen miles beyond Beaver Falls bothered me. I glanced into a pool roiling in turquoise and shook my concern. Reggae might last a few more years, and radioactivity many millennia longer. But the Canyon would endure long after all of us are dust.

"Today, about thirteen miles downstream, we'll be running Lava Falls," I had announced in my best Dr. Doom style before we launched this morning. I always did this. I had to.

"Many boatmen consider it the largest navigable rapid in North America," I had continued. "And even though we've run it many times, things can still happen there. We don't have it down pat. So, after we scout, make sure your ammo can is strapped down tight to the deck, your canteens are stowed in the stern where they won't wrap around an ankle, and that you don't have anything loose on the boat—especially stuff that dangles from a cord—and that your life jackets are snug. We have room now in food boxes to stash your extra stuff. Just ask. . . . And may the Force be with you."

Actually Lava was not that bad—depending, of course, on what part of the rapid one ran. It could be bad. It has been bad. It has been fatal. But reciting its credits backfires because many people think we are just hyping the place. By this point in the trip, they cannot believe that any of us actually worry about a rapid. They assume that Lava has become routine for us and that our war stories are hype. So we usually save those stories until after the rapid, except to joke. But not expounding on Lava in advance makes some people more nervous than they should be. Sometimes you can't win.

It depends on what happens in Lava.

Lava—not the rapid but the volcanic flows of basalt in the Canyon that cause the rapid—makes its first appearance three miles upstream from Lava Falls. Most people do not even notice the blocks of once-molten rock perched in the mouths of hanging canyons hundreds of feet above the river. But my first glimpse of these little chunks of columnar basalt always starts an emotional chain reaction—a Pavlovian response to Lava Falls. My heart rate increases several lub-dubs per minute, and I start imagining last-second moves.

This lava reveals that even though this western section of the Canyon may be only about 5 million years old, it's been geologically busy. The most important geological entity affecting the fate of this region—and that of river runners—is the Toroweap Fault, which crosses the river on a north-south line at Mile 179. Although the Toroweap Fault predates the river and the Canyon, it is responsible for the most recent and radical changes in both.

The Toroweap is a transverse fault, an interface between two chunks of crust moving against one another. The east chunk has been thrust upward 700 feet higher than the landscape immediately west. At this moment, rowing a couple of miles upstream from the rapid, our boat was floating level with friendly ledges of Tapeats continuing downstream to Lava Falls. Below Lava, the river flows against Muav Limestone suddenly 700 feet lower.

So what? Two things. First, erosion in this topographic maze we call Grand Canyon is most severe along weak areas created by faults. Flash floods from cloudbursts dutifully follow the "drawings" etched in rock by faults on the master blueprint of the landscape. Over the years, tributary canyons entrench along these because the zone of sheared, weakened, and weathered bedrock created by faults yields most quickly to the corrasive power of flash floods. Even before Grand Canyon was grand, the Toroweap Fault had already marked the course near Mile 179 of two huge tributary canyons: the one to the north is called Toroweap Valley, to the south, Prospect Canyon. They meet mouth to mouth at the river. At least they used to.

Today, and for the past million years or so, their intimacy has been shattered—actually, plugged. But before this happened, these twin tributary canyons set the stage for a humdinger of a rapid due to the doubled load of alluvial boulders they dumped into the river. So, a million years ago, before there was any lava in this region, there probably was a big rapid at Lava Falls, maybe bigger than today.

The most important consequence of the Toroweap Fault is not the erosion along its length dictating these tributary canyons. More important is its depth. It is so deep that it offered an early escape route for magma trapped beneath the crust here.

Not long before 1.2 million years ago, lava burst to the surface from a cluster of volcanoes nearby. Four cubic miles of molten rock flowed across the broad Esplanade atop the Supai, then poured down the canyons, ultimately filling Toroweap Valley, Prospect Canyon, and Grand Canyon itself at least 2,330 feet deep. Behind this dam of molten basalt (Prospect Dam), the Colorado roiled and steamed as it ponded in defeat to form a lake that may have extended all the way upstream to Mile Zero at Lee's Ferry. The dam was so huge that the river took twenty-three years to fill it. Many writers have described the scene that must have ensued here at that geological moment, but the first to try, John Wesley Powell (who portaged here), succeeded best: "What a conflict of water and fire there must have been here! Just imagine a river of molten rock running down into a river of melted snow. What a seething and boiling of the waters; what clouds of steam rolled into the heavens!"

Just imagine! Imagining, in this case, is a lot more convenient than witnessing. This conflict of water and fire, notes geologist Otis Willoughby, happened not just once but a dozen times, though no subsequent eruption produced a dam even half as high as Prospect Dam. At Mile 179½, packed against the north wall that used to be the deep canyon we call Toroweap Valley, are flow upon flow of basalt. In total, eight flows are juxtaposed unconformably in this one spot. All of them were belched from the earth via some weakness in the bedrock created by the Toroweap Fault. Some of these eruptions flowed down the Canyon and usurped the course of the river for more than eighty miles. After each eruption the river was dammed, with the last two such dams occurring about 640,000 and 140,000 years ago. During the years between flows, the river scoured away most or all of each lava dam, although it occasionally abandoned old sections of plugged riverbed (such as at Mile 183 north) by detouring around them and carving a new channel adjacent to it in the old sedimentary rock. These old lava fills in the riverbed reveal that the Canyon was cut to within fifty feet of its present depth a million years ago.

It is still hot beneath the rock here. Immediately south across the river from Lava Falls is a jungle of lush foliage growing on a broad shelf of travertine deposited by warm springs. It makes one wonder if the Toroweap Fault has delivered its last load of lava.

Despite more than a million years of severe erosion, this igneous superglue is still plugging both original tributary canyons on the Toroweap Fault as if it were intended by nature to repair an error in her erosional blueprint. But the upper drainage of Prospect Canyon was too extensive and entrenched by then to change its habits. All those flash floods had to go somewhere.

The last thousand millennia of flash floods from the southeast have simply bypassed the 2,000-foot-thick fill of lava in Prospect by veering a quarter mile southwest (downstream) and carving a new Prospect Canyon more than a mile long. Because this new Prospect Canyon is so large already, geologists were tipped off that the lavas filling the old one must be the earliest and oldest in this region. Be that as it may, the new Prospect Canyon, being steep and draining the large basin of its prevolcanic ancestor, has dumped a huge load of boulders into the river immediately downstream from the Toroweap Fault. These boulders create Lava Falls. And a lot of nervous sweat.

We floated passed Vulcan's Anvil, a solitary block of ebony basalt emerging almost forty feet from the river at Mile 178. Prickly pear cacti grow from several hairline fissures between its hexagonal columns—making this one of the world's tiniest inhabited desert islands. This island in the Colorado is a volcanic neck, the solidified interior of a volcano that erupted dead center in the river. Its source must have been a minor dike branching hundreds of yards upstream from the igneous pool feeding the volcanoes that poured the mountains of lava from each rim.

I stared at the polygonal columns packed into this alien-appearing knob. When basalt cools, it solidifies into packed polygons that switch geometrically from tetragons to hexagons as they cool and crystallize inwardly. These columns seem to be the result of intelligent construction—stacked raw materials left over from some megalithic age (hence earning such names as Devil's Postpile in California, and Giant's Causeway in Ireland). Downstream from here the flows packed against the walls of the Canyon stage the geological equivalent of an artist's showing—an incredible display of polygonal geometry. I have seen this same columnar jointing in river canyons in Africa, Asia Minor, Sumatra, and South America. Thick flows seem to develop thicker polygons than thin flows do. Also, the base of a flow, called the colonnade, forms much thicker columnar hexagons than the entablature of long, sinuous polygons radiating above it. Gazing at them half persuades me that some intelligent message lies hidden in their geometry.

I pulled my eyes away to glance downstream at a cinder cone perched at the brink of the left Esplanade above the mouth of Prospect Canyon and the still-hidden Lava Falls. My heart lub-dubbed a few beats quicker.

T. A., Fabry, Danny, and I perched on the huge block of basalt thirty feet above the tongue of the rapid and stared down at the all-too-familiar chaos of waves pounding over submerged boulders dumped here by those flash floods from the new Prospect Canyon. When I had first stood here to scout this complex rapid in 1976, the flow level had been higher than now, and I had seen no good route through it. The route I finally chose, in ignorance of where other professional guides were running here, was the route called the Dory Slot or the Right Slot. This route had worked even better than I had hoped. For my next dozen trips or so, I stuck to the Slot like glue and polished it like silver. I only started giving serious consideration to alternate routes two years later after a notorious Canyon boatman named Whale said to me, "You haven't run Lava until you've run the Right Side."

I had watched two boats run the Right Side on that first trip. The boatman in boat number one ran tandem with me on my first run. Only seconds behind me, he was blasted completely across his boat and into the river like a fly hit by the stream from a garden hose. His boat buckled in two. His aluminum rowing frame bent, then cracked into pieces. As he swam for his life I rowed upstream to intercept him and some of his flotsam. He had confirmed my initial suspicion: the Right Side was for idiots.

I then hiked back up to observe our third boat follow down the Right Side. Maybe, I thought, the trashing of that first boat on the Right Side had been a fluke. I watched. The boat slid off the first diagonal wave, dropped into the V-wave, then bent at right angles. The boatman was knocked off his rowing seat into the bilge in the stern—still gripping both oars by their handles. He struggled there like a turtle on his back, half drowning in two feet of bilge water for what seemed about a minute but was actually only five seconds. Meanwhile, his boat, swamped and out of control, caromed into even bigger waves. Ugly.

But Whale's pronouncement would not stop ringing in my ears, and eventually I arrived at Lava to find the water so low that

238

the Right Side was the only good option. But then it was easy. So I started pushing the envelope on subsequent trips. I studied every nuance in current roaring down that flume of insanity at higher flows, identifying within inches the ideal position for a boat at each critical juncture. Then I started running it.

Whale was right. It was wild. But as the flow levels increased, one's ability to predict one's position and momentum after emerging from the V-wave went down the toilet. I tried slight variations of entry, but at higher flows any finesse was canceled out by the funneling tendency of the V-wave. Still, my good runs far outnumbered my sloppy ones, and I had never gotten into serious danger here. On my last trip we had all run the Right Side and two of us had aced it. I had been half of them.

"I'm gonna run the Right," I announced matter-of-factly. "So I'd like someone to be down there first."

"The Slot's the run," T. A. responded. "It's clean. Why take chances?"

Danny, still uncommitted, stared at the rapid. The Slot *was* clean, really clean, almost boringly clean, but it required very close attention to navigation. Missing the entry by a foot to the left meant being sucked into the Center Trough and hurting your people. Missing it to the right would automatically funnel the boat into the Right Side but with the wrong momentum and with an almost total lack of control. Danny stared again at the Center Trough and shuddered slightly. The window of opportunity there was so narrow and his performance so far on this trip so sloppy that he decided that the Slot was too risky. And the Right Side was out of the question. "I'm going Left," he finally announced. The Left Side was choppy and sloppy, but since the flood of 1983 had cleared it, it offered a route even at medium flows.

"Who wants to run first?" I asked. We would run this in two groups so our passengers could photograph one another in big water.

"I'll go," T. A. volunteered.

"I'll go too," Danny said.

Fabry stared at the rapid silently, then said, "I'd like to run second."

I turned to Sutton and Jimbo, our trainees. "Et tu?"

Sutton pondered, then admitted, "Well, . . . I'd like to watch."

Twenty minutes later two boats appeared above the tongue. My heart rate sped up. T. A. ran the Slot and did a nice job of it. Danny ran the Left, and to the relief of his worried passengers, emerged unscathed.

"That didn't look bad," Rowan, one of my passengers, announced. He seemed disappointed. In contrast, his wife, Sheri, stood smiling like a lottery winner and seemed inordinately pleased about the ease of Lava Falls.

"We're going to run it a little differently," I admitted, "on the Right Side. We'll be going for the gusto."

"Good," Rowan approved, perking up. Don nodded in agreement. Tina stood with wrists on hips and head cocked as if to offer an opinion but said nothing. Sheri suddenly appeared crestfallen. Gusto was definitely not on her preferred list. Safety was.

We had stowed the last of our sundry impedimenta, tightened the last strap, and tested every grip line, and I had made sure that my oarlocks were tight in their aluminum sockets. I looked to Fabry.

"Ready?"

He stuck his thumb up and grinned. Jack and Lydia, two Brits sitting in the front of his boat, smiled with him.

"Can you see Jimbo?"

He stood from his rowing seat to peer over a block of basalt in the water, then nodded. "They're ready."

"Have a good run," I offered as I pulled away from shore. He was going for the Slot, Sutton and Jimbo for the Left. We would be the only ones down the Right Side.

"Have a good run, Michael," he replied. We both meant it.

As usual above major rapids here, the water was a calm antithesis to what awaited us below the tongue. Its calmness seemed almost accusatory. *Choosing* the most challenging route in Lava Falls was senseless bravado, this calm water seemed to whisper. I tried to ignore it.

"Okay," I started, "we're running the far Right. And, while I don't think we're going to eat it, if anything *does* happen and you find yourself outside the boat and in the river, swim *left* with all your might. Don't worry about anything but swimming left— till you are past that huge black rock at the foot of the rapid. Then swim to the right to avoid Son of Lava. Is that clear?"

Sheri and Rowan and Don and Tina were packed hip to hip in a solid phalanx on the front deck. They all nodded. Sheri looked as though she had just swallowed something that had changed its mind.

"Also, I'd like one of you guys to drop forward and wedge yourself in the nose to get your weight up front. Any volunteers?"

"I'll do it," Rowan said immediately. Unlike his wife, he was looking forward to this.

"Great. Thanks. But don't move forward until we're there and I call 'forward.' And as soon as you're there, grab the bowline and the grab line."

"Got you."

Why *was* I running the Right Side? I asked myself again. Was it because this would be a lot more fun, . . . because it was a challenge, . . . because Lava Falls was the last really big rapid of the trip? At this water level, 13,000 cfs, it was the biggest rapid on the entire river. Or was it because you haven't rowed Lava until you've rowed the Right Side? Or was it because this might be my last trip (you never know). I knew that this was probably the only trip these people would make here—until their kids grew up.

But was I risking making those kids orphans by running the Right Side? Or my own kids orphans? People had drowned here. I hated this calm float. Why did it have to take so long? But all of those fatalities had been passengers, passengers on motor rigs.

No professional Grand Canyon oarsman had been killed here or had lost anyone here. So was I about to make history? What a stupid thought. I reviewed my plan of action and felt my heart rate soar in response to the mental images.

This is ridiculous. I had it easy. I had a top-notch boat under me, tuned to perfection. I knew the route better than most people know their own living rooms. Not everyone had had all of these advantages here.

I thought about James White. He had been dragged sunburned, emaciated, and exhausted from a driftwood raft at Callville, Nevada, on September 8, 1867 (two years before John Wesley Powell's expedition), after what he claimed had been a ten-day float through the most horrendous river canyons known to man. He had been prospecting with two partners in the San Juan Mountains when they had been attacked by Indians and forced to abandon their horses to flee down a tributary canyon of the Colorado. One partner, Captain Baker, was killed by Indians, the second, George Strole, drowned on day four. Two rafts self-destructed, White claimed; the one the friendly Mormons found him clinging to had been his third. Although White claimed for the rest of his long life that he had only run one really bad rapid, a botanist named C. C. Parry, who was attached to a Union Pacific survey party, convinced both White and himself—and the readers of his article in the *Transactions of the St. Louis Academy of Science*—that White had floated the entire Grand Canyon plus Glen Canyon and then some while roped to three or four cottonwood logs.

Had James White really beat Powell by two years? And had he floated through Lava Falls willy-nilly at a flow matching almost exactly this one? Few people think he did—not because it is impossible but because when White described the canyons and the river to Robert Brewster Stanton forty years after the journey, he did not describe Grand Canyon or the Colorado as it is here. So where *had* he been for ten days floating at three miles per hour, fourteen hours per day, a total of 420 miles? From the San

Juan River to Callville along the Colorado was roughly 450 miles. Maybe the poor wretch did run Lava on that driftwood raft. Maybe Lava had been his one really bad rapid. If so, what did I have to worry about?

Plenty. But even if White had spent most of his ten days swirling around in eddies and making very slow headway after having launched somewhere on the river at the foot of Grand Canyon a hundred miles downstream from Lava Falls at Pierce Ferry (although how would he have gotten there from the San Juan Mountains?), I had my hands full.

But still it was relatively easy. On Easter Sunday in 1955, Bill Beer and John Daggett arrived at Lee's Ferry to jump into one of the dumbest adventures in the annals of the West. They donned wool long johns, neoprene jackets, life vests, and swim fins, and then grabbed their two big rubber bags of provisions and supplies each and waded into the Colorado. They were planning on *swimming* it, all the way to Pierce Ferry (Mile 280). And they did—twenty-six days of near hypothermia. Here at Lava Falls they had no boat, no cottonwood raft, and not much hope either. Staring at this rapid had almost paralyzed them.

"How does it look to you from here?" John shouted to Bill while scouting.

"Worse!"

"Think there's a chance, Bill?"

"What?"

"I say, do you think there's a chance we can swim it?"

"No."

But they had, although mostly underwater. And they survived it. They even had fun.

So why worry? Because there were two little kids waiting in Flagstaff for their father, a father trying to extract the last, and the ultimate, essence from this river before losing it.

Suddenly the horizon of Lava Falls rose into view. I had photographed the entire rapid in my mind—a good thing. Because of its drop (reputedly thirty-seven feet but actually only thirteen), we could see none of Lava Falls from here. Now I had to line up for one of the three entries. In seconds changing my mind would be stupid—or worse.

Pivoting the boat slightly, I lined us up on the left diagonal of the far right tongue. A precision entry was critical. We would glide down this long wave, then slide abruptly over the sharp left diagonal at its end. Once over that, I would plant my right oar and pivot the boat nearly ninety degrees, on a dime, to hit the approaching right diagonal straight on. Once over that wave, we would funnel into the V-wave dead straight. We *had* to hit it straight.

As I took the first critical stroke, I felt that familiar time-warp sensation. Adrenaline was kicking in, in a big way, and despite all of these landmarks being invisible until the last second, I knew that I could hit each one as accurately as a surveyor with his transit. It did not worry me. I could lead this dance with Lava right up to the V-wave. It was what might happen in the V-wave—and immediately afterward—that really worried me.

"Okay, Rowan, get forward!"

A space suddenly gaped on the front deck. Hips slid together to fill it.

"Okay, hold on!"

Splash. The Domar bucked over the lower left diagonal. I pivoted. We splashed harder into the right diagonal. We were on track. The V-wave yawned before us like the entrance to a tunnel of love that had suddenly switched to the dark side.

Why the Right Side? echoed again in the back of my head. Shut up.

"Lean forward and hold on!" I hunched forward on the rowing seat, pressed my calves against the side boxes to wedge myself, and held my oar handles in a death grip.

Sploosh! We all vanished under water. "Come out straight," I commanded the boat.

"Oh shit!" I heard myself swear. The V-wave had belched us out toward the right shore. Even worse, it had shoved us there stern first. To get us out of trouble, I had to row us left. Dead ahead was the huge, infamous black boulder of lava, Dead Man's Rock, interrupting the flow on which we were now trapped. That rock had destroyed several boats beyond recognition. In five seconds we would be there. But with my stern to the right wall I could not push this boat swamped with two tons of water to the left. Nor did I have time to pivot the necessary 180 degrees. Polished columns of basalt raced past us almost within touching distance. The lava monolith ahead grew as if someone were magnifying it with a zoom lens. I pushed on my oars to the left and felt a surge shove us back even farther right.

When Sheri heard "Oh shit!" she knew it was over. This was not her first river; Rowan and she had done many. But their last experience on the Youghiogheny had almost killed her. They had lain awake most of the night before for ten rainy hours in a leaky tent. Because the rain never stopped, they expected the outfitter to cancel and give them a rain check. But no, the trip was on. Two couples to a boat—and with no guide on board. Directions at critical rapids would be yelled to them. After hours of drizzle they entered the most critical rapid, zigged when they should have zagged, hit a rock, flipped their boat, and all four of them had ended up in the dark, icy womb of a giant washing machine under the boat. Sheri was five months' pregnant. The wild water and darkness became a nightmare. She hit bottom and swallowed a week's ration of water. Before Rowan finally shoved the boat aside, Sheri had resigned herself to dying with her unborn child. Eventually she felt Rowan tugging her to shore. And she thought she might live. But with no sun they got colder and colder until hypothermia set in. It had been a nightmare. But it had been small potatoes on a trivial little river. *This* was the Right Side of Lava Falls on the mighty Colorado . . . and the guide was saying, "Oh shit!"

But everyone else was having a great time. Ignorance is bliss. It was wild out here, like jumping into the screen during a disaster movie.

Almost everyone else. I was not happy with the current turn of events either. I shoved again on the oars, hoping against hope for lateral gravity slop to edge us left. Grasping at straws. I pulled on the right oar to straighten us for the inevitable collision with that black monolith that had caught hundreds and hundreds of boats, absolutely destroying some of them.

En route, we plunged through two more huge waves and troughs, each of them burying us so deep that we had to hold our breath.

The black rock ahead now seemed the right size for King Kong's throne.

"High side and hold on!" I yelled.

Feverishly, I pulled one last stroke on the right oar, pushed on the left, and got the boat to straighten out. The mammoth collision of the river with that rock surged us up so high that, for an instant, I could see over the top of it to the other boats waiting in the eddy below. I saw eyes in circles of surprise.

Miraculously, the surge subsided and we went with it. I never even felt the slight bump that would have come had the boat touched the rock. The current swished us around the monolith and into the haystacks of the tail waves. Relief washed over me like nothing in Lava Falls had. "Whew," I heard myself utter as I pivoted to pull into that eddy. Luck of the Irish.

Close. Very close.

Everyone spun on the front deck and cheered. No one who has ever won an Academy Award could have faked greater elation. Rowan effused, "That was a *great* ride! A *great* ride!"

I smiled weakly. It was as if the hangman coming for me had just tripped, stumbled, fallen through his own trap door, and hanged himself.

Rowan asked, "Can we go back and do it again?"

The answer to the question, Why the Right Side of Lava? was pretty obvious now.

Because it was fun.

16
Golf Ball Beach

I should have seen it coming. It started with lasagna. Postadrenaline lasagna. Innocent enough. Our deliverance from the Right Side of Lava was now far enough in the past to take it for granted. And to start cooking. Sutton and I were about to tackle our most time-consuming meal.

For Post-Lava Night on this trip I had chosen a rare type of camp: wind-protected and with early shade. At 3:00 P.M., after celebrating below Lava, we had arrived here on the north side near Mile 183. Two hundred yards of huge crescent beach nestled against a lava cliff bordering a cove with a half-moon eddy so shallow that we could walk out into it for twenty yards. Here in the land of lava the view was unearthly. Massive flows of black lava—now frozen as if in a Polaroid snapshot—had poured into the Canyon a dozen times in the past 1.5 million years from above the Esplanade. This stark region seemed an unfinished piece of cosmic landscape, a cross between the American Southwest and the Moon. Although camping here would force us to row thirty-seven miles tomorrow, it was too perfect to pass up. But I should have.

Before tackling the lasagna I had hiked upstream beyond the beach to find a micro-eddy for bathing beyond view. When I came back, I found one of our passengers, Chantal, lying along the shore, the lower half of her body in the fifty-five-degree water, the upper half on the sand, her eyes closed.

Minutes later, Sutton, our Cinderella on this trip, and I buzzed within the defensive perimeter of the kitchen to get the lasagna rolling. Now the wash pails were heating, and a big pot had about ten diced onions and some chopped garlic sizzling in it. Sutton cranked the can opener to open the four big cans of stewed tomatoes, three of tomato sauce, six small ones of paste, four more of mushrooms, plus two of spinach, to dump into the fourth, even bigger pot to start the sauce heating. But something was missing.

I looked around. All of our menus and recipes were in my head, but not all of the ingredients would list in my mind on command. I stopped to try to recall what was missing and gazed around for inspiration. I spotted T.A., Fabry, and Danny down near the boats with three golf clubs and a bag of balls. T. A. was teaching Fabry how to drive a ball. I automatically scanned the shoreline upstream to make sure no one was present on their improvised fairway. The beach was clear.

Italian sausage. Of course. In my cooler.

I skidded down the dune to the bowlines and leapt onto the bow, enjoying the bounce. Lava day. Then I dug into the meat cooler for the vital ingredient to make our lasagna rise above that of the competition.

More sizzling as I squeezed the disgusting innards of the sausage onto the browning onions and garlic. On top of this I piled a heap of oregano, a much bigger heap of basil, a secret pinch of chili powder, and some bay leaves. Eye of newt and toe of frog, wool of bat and tongue of dog—

T.A. stood over the golf tee, swiveled his hips slightly, swung, and drove a ball down the beach. He turned away from admiring its trajectory and said, "Okay, Fabulous," offering the club to Fabry.

248

Fabry stood over the tee, did not swivel quite the same way, and took a rigid swing. The ball limped into the sand.

"No, no," Danny admonished impatiently. "Here, let me show you."

Fabry reluctantly relinquished the club to Danny. He placed the ball on the small green that T.A. had brought, stood up, concentrated, and swung as if trying to decapitate a striking cobra. He chipped the ball, spinning it into the sand a few feet away.

"Nice," Fabry observed.

"Wait, let me try it again," Danny protested in self-annoyance.

T. A. handed him another ball.

More careful now, Danny swung again and connected solidly with the ball, sending it in a flat trajectory about a hundred yards along the beach, but instead of flying parallel to the river, the ball disappeared up toward the cliff, then bounced at a right angle off something hidden and flew back toward the river.

"Adder's fork and blind worm's sting, lizard's leg and howlet's wing—" The cheese—hard to forget, but I had. Back down to the boats and the dairy cooler.

Still a lot of cheese in here and only forty hours left in the trip. I dug for the four-pound block of mozzarella and the huge tubs of ricotta, thinking that in forty hours I would get to hug and kiss Cliff and Crystal and Connie. "You should have seen my Right run in Lava, Darling . . . "

Danny bounded onto the boat next to me and feverishly dug into a cooler. "I screwed up, Doc," he admitted, his face contorted into a study in anxiety.

"What do you mean?" I asked, suddenly realizing as he pulled the dripping remnant of a block of ice from the cooler what must have happened but hoping that I was wrong.

"I hit somebody with the ball."

"Somebody? Who?" Then I knew on my own. "Chantal?"

"Yeah."

Normally Danny was glad to be alive. Before hitting twenty-one he had fought and won a long battle against a cancer that his doctors were sure would kill him. They were almost right. And Danny was extra happy on this trip because—despite not being on our guide schedule at all—this was the second trip in a row on which he had been asked to work as a last-minute replacement. Hollywood casting could not have found a better rookie: Danny was young and tall and thin. His windblown hair always looked like he had forgotten he owned it, and his face seemed designed to express amazement. Being alive after every authority expected him to die led Danny to live for today, to extract every possible ounce of gusto from the moment.

But on this trip he had gotten over his head, not just in white water, in camp. It had started as a prank. Unknown to me, Danny imagined that one of our passengers desired him—despite her having arrived on the trip with a date. Chantal was forty, but her lean California Girl body and suntan made her appear fifteen years younger. At Chuar Canyon on the fourth evening—the evening prior to our entire charter of clients departing the trip at Phantom Ranch—Danny was pushed into making his move.

"Hey, it's your last chance," T. A. had prodded Danny.

"But she's in the tent with Mark (her date)!" His words slurred from beers too numerous to remember.

"So? Think of something."

"Maybe I can get her to come out."

By this point I had begun to suspect that I had missed an important development while leading all the hikes, ones that Chantal (and Danny) had avoided.

Danny leapt off the boat. Mark and Chantal had pitched their tent nearby but deep in a cluster of gnarled mesquites through which the moonlight barely filtered. Danny wormed his way through the maze of thorny branches to the rear of the tent. Once there he hesitated, unsure how to invite Chantal out while not letting Mark know that she had gone. He finally decided to wing it.

250

He slithered around the side of the dome tent and scratched at the fabric near the door, trying to emulate the probings of some insignificant but irritating nocturnal creature.

Nothing happened.

He scratched again, harder.

He heard rustling and quickly backed up to conceal himself behind the tent. The zipper whined and a flashlight beam shot out to illuminate the sand, boulders, and mesquites. He peeked at who was holding it.

Mark.

But Mark seemed unwilling to leave the tent to search for the creature who had been scratching.

Danny squatted there and realized that his plan was not working. He did not want to return in defeat to the boats, but no other option occurred to him.

"But Danny, you're not going to let a little thing like that stop you, are you?" Fabry had asked, almost rhetorically. "This is your last chance."

I suddenly felt sorry for Danny. T. A. and Fabry were leading him down the mesquite path.

Danny leapt off the boat again and stalked into the darkness. This time noisier. His logic now was that if Mark had gotten up the first time to look for the scratching, then maybe Chantal would look the second time.

He ran his fingernail along the nylon. Again he heard rustling, the whine of the zipper, and was half blinded by a flashlight beam darting across the ground in front of the tent. But again Mark held the light.

This was definitely not working, but desperate situations require desperate measures. Danny waited for everything to calm down within the tent, then, he told us later, he crept to the front.

Working the zipper as quietly as possible, he opened the door on Chantal's side of the tent enough to reach inside. He then found what must have been her leg and scratched it. No one, he thought, could sleep through this.

But the leg may as well have been attached to a corpse. No one in the tent twitched.

Far less stealthily than he had crept up to the tent, Danny now admitted defeat and stumbled back to the boats.

"So what happened?" Fabry asked.

"I couldn't get her to wake up!" he said, sadly.

"Hey, it's your last chance," T. A. reminded him again.

But it had not been. The next morning T. A. persuaded Chantal that Grand Canyon downstream from Phantom Ranch was even more beautiful than what she was seeing here, adding that we had one open space left on the boats. She decided to abandon Mark and to continue downriver with us—and at the same time tried to pay for the second half of the trip on Mark's credit card. Danny knew that she had stayed to be with him.

But she had repulsed his clumsy advances on the first couple of nights below Phantom. Then, paradoxically, she had whispered to him at lunch of day seven something like, "When a girl says no, it doesn't necessarily mean no forever." This spurred a torrid river romance.

"Well," I asked, "how is she? Where did the ball hit her?"

"Bad. It hit her in the head."

"*How* bad?" I asked, mentally preparing myself to go examine her. I was an EMT (Emergency Medical Technician). So was T. A., and we had two M.D.'s on the trip: Rowan was a pediatrician and Jack an anesthesiologist. But none of us could do much with a real emergency down here.

"Bad."

"Is she conscious?"

"Yeah. T. A. and Fabry are with her now."

"Okay, let me go get Rowan and Jack." Before I did that I went to my boat and pulled out the radio. I still was hoping we would not need it.

Despite the early shade, the temperature was still at least ninety degrees on the beach, but Chantal was shivering violently inside her sleeping bag. Danny had created an ice pack with a shirt, and they were holding it against a huge goose egg on Chantal's right temple. Already it was turning black. She looked like she had been dragged out of a head-on collision.

It was easier now to understand how no one had known she was here. Her camp spot was the only one on the upstream half of the beach, and it was hidden high up toward the cliff in a depression, a stream channel that had eroded years ago, probably in 1983, due to a reverse eddy in the main eddy at flood stage. This little dip was also concealed partly by tamarisks but not enough to filter out a line-drive golf ball.

"Let me take a look at her," Rowan said, taking control. Jack squatted down with him.

T. A. and Fabry gave up their patient to Rowan and Jack. The three of us quickly moved away to palaver.

"We've gotta get her outta here," T. A. started. "It's ridiculous for her to stay down here."

Fabry nodded, "Yeah, we should get on the horn and call in a bird. She could have complications. It's not worth the risk."

I agreed, in principle. I had been here before—more than once. "I agree, but I'd like to hear what the medical boys have to say. So let's hold off on the radio until we get their assessment."

"They don't *know*," T. A. objected. "What are we gonna do if she has a concussion? We should start tryin' to get someone now."

"Just hold off for a couple of minutes."

"Well, what do you think?" I asked as Rowan and Jack came over.

"She's not in danger," Rowan answered.

"You're sure?" I asked. "That's your medical consensus?" I added, looking at Jack. "What about the possibility of a concussion?"

Both men nodded. "She's got a good bump, but I don't think a concussion is likely," Rowan said. "Neurologically she appears okay."

"I'm thinking about evacuating her," I admitted. I had been in this situation twice before. Both would have been fatal.

The first time had been at Lava Falls during the flood of 1983 at 90,000 cfs. The rapid had been the worst ever—horrible. Survival demanded a desperate effort to break out of the 200-yard-long tongue on the far left to avoid being funneled into the center and into the gigantic reversal of the eddy fence there that flipped boats our size as if they were pennies.

Alister Bleifus and I had run it first, more or less as guinea pigs. The power of the current rushing right and the mountainous diagonal waves had made us feel almost ineffectual—but not quite. Both of us had rowed our guts out to make it, then continued through Son of Lava to pull over at the earliest opportunity on opposite sides of the spate. Then, from a third of a mile away, we had watched the next three boats. It had been hard to see them because of the fifteen- and twenty-foot waves hiding the relatively dinky Domars, but suddenly I spotted a shiny, smooth floor atop one of the tail waves.

"Flip!" I had yelled across the river to Alister.

The swimmers were still so far upstream that neither of us made a move. We would have to wait and time our exits to intersect the courses of the five people in the drink. Already our other two boats were in the river shortly ahead of them. If they slowed down, they might rescue most of the swimmers.

Craig, the boatman who had been rowing, now tried to climb onto the floor of his flipped boat to drag his four passengers to safety. He could not manage it. Shouting directions to the others, he tried again, this time attempting to grab the bowline and toss it the full length of the floor. But it eluded him. Meanwhile the current rushed them into the curving Son of Lava, where mountains of water slammed into a cliff of Muav.

The upside-down boat rode through the waves toward the cliff. Some of the five held onto it, others were swimming frantically to the right to escape the wall. One woman, an Australian, clung to the boat but ended up between it and the cliff, where her head was slammed between them. The Domar then bounced away and she became separated from it.

I saw her coming. I waited, then pulled out. The trainee boat, however, was closer and picked her up before she could drift down to our position. By that time everyone else, plus the boat, had been nabbed by one or another of us.

At lunch she was fine, as plucky as Cactus Annie. She joked about the flip, about how cold the water was, about getting dragged out. But an hour after the accident her comments stopped making sense. A quarter hour later she quit making comments. Another quarter hour and she was comatose. She had suffered a concussion, and the slow hemorrhage and swelling in her cranium had pinched the arterial feed of oxygenated blood to her brain. She was in trouble.

The weather was cloudy and windy—terrible for a chopper—but we had no choice; we had to get our message out. Our radios are line-of-sight. One must *see* a receiver because broadcasting blind runs the risk of killing the battery and destroying one's sole ace in an otherwise bad hand.

We camped immediately and pulled out the radio and signal mirrors and stared into the sky. Scenic flights were common near Lava Falls but not in weather like this. Finally we contacted a commercial airliner at high altitude en route to Los Angeles. The pilot relayed our SOS and position to park headquarters.

Despite the wind, the chopper came in, waited for a hole in the weather, then lifted her out to Flagstaff Medical Center. Two days later she was okay.

This was a strong lesson to me in how sneaky a head injury can be. Sneaky is the rule for concussions, not the exception. Concussions normally develop so slowly that the victim seems okay—in pain maybe, but lucid—for about an hour. By then the epidural hematoma or subdural hematoma has caused such internal swelling that sections of the brain become oxygen-starved and close up shop. While I was impressed by this lesson, I had hoped that I would never need to act on it. Three years later, however, I was put to the same test.

This time we were hiking up Tapeats Creek on the way to Thunder Spring east of Surprise Valley, perhaps the most beautiful hike in the Canyon and one that followed a well-beaten trail. Because Tapeats was in spate, I had rushed ahead to ford the current and fix a rope as protection when the main group showed up. Just as I was about to tie the end of the rope on the far side, a passenger rushed up and yelled to me that Dusty had fallen down to the creek. I struggled back across Tapeats, imagining that she had fallen *into* the creek in a narrow section and now was bouncing downstream in the miniature but powerful white water. But when I reached the scene I saw that Dusty had slipped off the trail and had tumbled nearly twenty-five vertical feet onto a boulder pile next to the creek. The man who had been walking behind her had witnessed her glance up, away from the trail, perhaps at the cliff. At that instant her foot slipped on fine sand atop a boulder embedded in the trail, and she slid off down the steep slope, nearly a cliff, separating the trail from the creek. Young and agile, she had spun catlike to grab for some handhold on the trail, but there had been nothing to grab. Her momentum carried her off the edge. She tumbled in two reverse somersaults down the embankment and hit her head at the base of her cranium (housing the "old brain," which is responsible for autonomic processes) against a pointed boulder at the bottom.

Pushing through the circle of people, including my four fellow guides and two M.D.'s, I sized up the scene. I felt like a Johnny-come-lately, but I had to know how she was. I asked her. She said she was okay . . . sort of. I noted the location of her head injury, now bandaged, and the multiple contusions and abrasions all over her, then looked at the blood-stained rock. My heart sank. Hitting that part of her head as hard as she did on a pointed boulder like this . . . I felt instant alarm. If I could have called in a helicopter that very second, it would have been too slow for me.

I pulled both doctors aside and asked them for their findings. One was in family practice, the other a neurologist. Both, however, were here together in a special subgroup of our clientele who were attending a wilderness therapy session aimed at ameliorating psychological problems. Their own psychological problems. And while they probably were competent doctors, I thought to myself, this was not the office.

"She's going to be in a lot of discomfort because of her multiple injuries," the neurologist said. "But I don't foresee any serious problem."

"She did not lose consciousness when she hit her head," the other added. "In ninety-nine percent of such cases, serious concussion would be ruled out."

For a moment I was silent. I was shocked, really. If I had ever seen a case of potentially serious head injury, this was it. It.

I thanked them and called a quick meeting of the guides. "I think she's in serious trouble; I think she's going to develop a concussion," I stated. "Once she's unconscious, we won't be able to do *anything* for her. I want to get her to where we can evac her as soon as possible."

"But the docs here say she's not in danger," Bart argued. "She needs to sit down in a quiet place out of the sun to rest. A little rest will—"

"She's not *tired*," I stated. "She has a serious head injury."

"But the doctors don't think she's in trouble," Stanley repeated the same argument. "I think we ought to give her a chance to rest."

Alister was also on this trip. A good thing. We had worked together on the Australian woman from Lava Falls. I caught his eye. He remained silent, but from the years I'd known him I could see he was troubled by what he was hearing from everyone.

"Okay, help her into the shade," I capitulated for the moment. Then I pulled Alister aside.

"You know what's going to happen to her," I said.

He nodded.

"How do you feel about beating feet back to the boats and getting on the radio?"

"I think it's a good idea," he said in his Missouri drawl.

"Good. Thanks. The radio is in my upstream side box. We'll start carrying her back to the beach in a few minutes."

Alister shouldered his pack, spun on his heel, and ran back toward the river, well over a mile away. At that moment I loved him.

I found Dusty's husband, John, and privately explained my fears to him.

"No, she's okay," he insisted. "She may not look it (she weighed only ninety-three pounds), but she's a tough girl. She'll be all right. The doctors said—"

I re-explained my fears to John, but he remained unconvinced. I went to talk to Dusty in her shady spot. With her were Bart, Stan, and Renee.

"Dusty, where do you live?" I asked.

She told me her address.

I turned to the other guides and whispered, "*Remember* what she tells us, because when we ask her again later . . ." Then I asked her for her birthday, phone number, age, why she was here, and so on.

"Okay," I told the guides privately, "I don't care what the doctors say. She's in big trouble and we're getting her out of here. Bart, you and I will carry her back. Stan (who had injured his shoulder in Crystal) and Renee, take the hikers on a slow walk back." Then I explained to Dusty that we were going to carry her back to the boats.

"Okay," she answered distantly.

No mutiny. And although I planned to carry her in shifts, Bart insisted on carrying her the entire way piggyback in a poncho rigged and slung against his back. He talked to her, eventually getting around to asking her the questions I had asked her earlier. Soon she could no longer give the right answers. Then

258

no answer that made sense at all. Before he started down the steep switchback trail leading to the confluence of Tapeats and the Colorado, Dusty was unconscious.

He set her down under a shade tarp we had rigged on the beach, and she went into a grand mal seizure.

"Any luck?" I asked Alister.

"Yeah, I got a message through about twenty minutes ago."

"Good work. So maybe an hour from now?"

"That's about right."

Despite her condition, Dusty's husband was still hoping she would be all right—even after the pilot landed the 'chopper on the beach and Ernie, the park paramedic, hopped out of it in his jumpsuit with his trauma box and IV setup. By now both M.D.'s were silent.

Except for Alister's help, this incident had felt to me like driving the wrong direction against heavy traffic on a one-way street. Ironically, for me Dusty had been the biggest ray of sunshine among the twenty-four of us. I had hated what had happened to her, and I too was hoping that her injuries were an inconvenience rather than a threat to her life. But I knew better, and after the chopper disappeared, I could not help wondering how she was.

Several days later, Flagstaff Medical Center released her. She told me a month later that she felt for weeks as though she were in the Twilight Zone. Eventually she recovered except for two side effects: dizziness upon quick movement and a total loss of her sense of smell. Her olfactory nerve must have been damaged. I still hope that this is not permanent. She and her husband own a fine restaurant. It must be tough to be unable to taste the food. Dusty sent me a card thanking all of us for our prompt medical help.

And now? Twice burned a fool. Despite my impression that Rowan and Jack were right about Chantal, I gave T. A. the high sign to signal SOS via the radio frequencies for western Grand Canyon. Then I returned to the defensive perimeter to boot the lasagna back into action.

Dana Morris, the park pilot who could land safely on the head of a pin during a cyclone, set his machine down on the beach at Mile 183 nearly an hour and a half later. By then Chantal had improved. Ernie jumped out with his trauma box and trotted over to examine our victim.

"How did this happen?" he asked as if in disbelief. Then he looked at Danny and shook his head. Ernie had a son nearly Danny's age, and the two knew one another well. Over the years they had gotten into situations together that Ernie had frowned upon enough to call the police and have them arrested. Small world. "A golf ball?"

Danny scuffed his foot in the sand.

Ernie quickly assessed Chantal and decided that she was not in serious danger, but he told her, "We are legally required to evacuate a victim when called in like this."

Meanwhile I lined Dana up with a friend who could line him up on a Bio Bio River trip in Chile, then I pulled out bread and peanut butter and jelly so that Ernie and he could make sandwiches. They had been working straight through the day and were ravenous.

Danny and Ernie helped Chantal to the chopper. She insisted on writing a deposition in fluent legalese exonerating Danny from any guilt in regard to her accident. Rather than seeming upset about the accident and evacuation, she was beaming. Almost beatific. She seemed as euphoric as a beauty queen who has just won the contest and is being crowned with a tiara and handed two dozen long-stemmed red roses. I did not understand it.

The chopper blasted sand into a local storm, then rose almost magically against the Canyon walls. Soon it vanished. We heard its rotor noise for a few more seconds, then only the river. The whole episode began to seem surreal. Strangely, at dinner no passenger even mentioned it. But Chantal's fate had infected my crew with the virus of superstition.

"I think she's gone to the dark side," Fabry said to me as we stood brushing our teeth on the boats in the dark. "There's something not right about her that I just don't understand." He paused. "I think she's a witch."

"Oh, come on, Fabry," I said. "I don't believe in that stuff—"

"Yeah, but look at her," he countered. "What are the odds that that golf ball would hit her in the head. Realistically, what? One in a million. She was invisible. Arnold Palmer couldn't have hit her with a hundred strokes."

"But—"

"No, I don't believe in that stuff either," he admitted incongruously, "But how do you explain *everything*. Remember Stone Creek Grotto?"

He was referring to four days earlier when we had spent the heat of the day in a narrow gash a hundred feet deep in the Shinumo Quartzite about three miles up Stone Creek Canyon from the Colorado (Mile 132) where cool water thundered over a waterfall. That afternoon all five of us had watched T. A. lead the hikers on the descent before we followed. Chantal had climbed down last in line. One by one they vanished among the redbuds beyond the lower waterfall, then behind huge clumps of bear grass, and finally amidst the catclaws. Except Chantal.

"So you think she's possessed, eh?" Jimbo had asked Danny—casually, as if he were asking if that six-pack was now completely gone or if one can still remained.

"She's into something . . . like she's gone to the dark side," Danny admitted, impressing himself more than any of us. "Maybe she's a witch."

Sutton and Jimbo both snorted at this. Fabry jumped in with a shield of logic: "If she was a witch, she could hear every word we were saying."

Danny started to nod in defeat at Fabry's logic but stopped cold as Chantal halted abruptly over a hundred yards away—beyond the roar of the falls—then turned back to face the narrows here and smile at us radiantly.

"And remember those ravens circling her at Havasu?" Fabry pressed me. "They were just *sitting* there. I mean, did you ever look into her eyes? They could be five hundred years old. All this stuff about reincarnation. She says she knew Danny in Atlantis!" He paused. "This is not your normal gal."

No, not normal. But neither was she a witch from Atlantis. Instead she was a victim of a freak accident. Maybe a bit too freak.

Dawn painted the walls a cool gray and then hot pink. It was 5:00 A.M. and Fabry had already gotten up to start the stoves and make coffee. We had a long day ahead of us on the river. I started my stretches.

Danny rolled over on his back deck and sat up. This was unusual. Normally he was dead to the world until the last possible moment forced him to get up.

"Doc, I just had this weird dream," he said out of the blue.

"Oh yeah?" I said. "What kind of dream?"

"It was like yesterday. Only instead of hitting her with a golf ball, I shot her in the head with a two-seventy."

"And what happened?"

"She sat up like this, with her hand on her head, and said, 'Danny, why did you shoot me?'"

Although I was there and saw it all develop, from Danny's moonlit crawl through the thorny mesquites a week earlier to the wild stroke of the iron last night, I still find it hard to believe.

The miles in western Grand Canyon always flow past easily. Here the river we float on is a peak release from Glen Canyon Dam nearly forty-eight hours earlier. The gradient is good and the current swift. Although in sunlight the temperature was about 110° F, out here on the cold river it was pleasant.

Here the wide, soaring, and torn cliffs of the Paleozoic sequence support black benches of basalt—survivors of a million years of river erosion after the volcanoes around Lava Falls and Whitmore Wash (Mile 188) erupted. They create an endlessly varied scenery. This is the lower Sonoran Desert. Ocotillo and cholla cactus dot the heat-baked landscape like statues on Easter Island. Desert bighorn sheep have staged a comeback here after the feral burros were removed in the early 1980s. They step delicately in this land of heat and rock, pausing to nibble the leaves of catclaw acacia or to butt a barrel cactus (which may have survived for a century) into the dust to gouge it open and eat the succulent pulp within. Despite the heat, the place seems friendly, hospitable, unthreatening. Making miles here is so easy that my mind often wanders somewhere else.

This happens to me here because the challenges of the trip are normally behind us by now. The river here does hold rapids, and I have witnessed flips on commercial trips and other sorts of mayhem down here. But I normally expect no more problems once we get below Lava Falls. Obviously, this is not always realistic, and with a crew half-spooked by black magic I had to remind myself to stay on guard to anticipate the improbable in time to nip it in the bud.

Even so, I knew I would see Connie and Cliff and Crystal tomorrow, and I let myself miss them as much as I actually did miss them. In the meantime, as the stupendous scenery of the lower Canyon drifted past, I daydreamed about new projects: books and screenplays to write, documentaries to film, research to conduct in the wilds again. My imagination galloped from idea to idea.

Spending twelve days on the river does this to me. Of course, this trip does not have to be thirteen days long. Thirty would be better, but I have rowed sixty-five miles in a day here and the whole 226-mile trip in five. But even five days is slothful compared to the record run.

The fastest rowing trip down here was run in June 1983. It resulted from the Bureau of Reclamation's premature filling of Lake Powell before the inflow from spring melt-off had peaked, which forced them to release more than double the previous record post-dam high. Three boatmen working for Grand Canyon Dories saw the bureau's mistake as the once-in-a-lifetime opportunity to set the record for the fastest human-powered traverse down the river. Kenton Grua, Rudi Petschek, and Steve Reynolds applied to the national park for a special dory training permit to run a quick, one-boat rowing trip, intended to beat their old speed record of forty-eight hours.

Of course, the Park Service being the bureaucracy it is, no one but the superintendent, Dick Marks, could make such a weighty decision, and true to his bureaucratic calling, he sat on their request.

To set a solid speed record, Rudi, Kenton, and Steve not only had to run on high water—the river was flowing at about 72,000 cfs then—they also had to row around the clock. This meant launching at full moon. But when that night came and went and then another and Marks *still* would not decide, their unique window of opportunity to push the envelope of rowing performance started to close forever.

By coincidence, I was starting a trip at Lee's Ferry when that window was about to slam shut. An hour before midnight June 25, 1983, I was nodding off to sleep on my rear deck when a truck roared up to the boat ramp. I heard a splash. I looked up to see by moonlight a single dory agitating in the water about thirty feet away with three guys in it fumbling to get their oars ready and whispering frantically to one another as if rehearsing for a Cheech and Chong movie. You can guess who it was.

And so could the National Park Service when a private boater who had also been on the ramp that night, a guy who probably had been forced to wait years to get his private permit, indignantly told a ranger the next morning that he had seen a dory sneak onto the river. By that time the outlaw dory was probably about eighty miles downstream and still rowing like hell. Around midmorning Rudi, Kenton, and Steve made a brief appearance at Crystal Rapid at Mile 98$^+$. Kenton later told me that they spotted a ranger on shore and knew they could not risk stopping to scout. Immediately the other two elected Kenton as oarsman. None of them had ever seen the rapid at water that high, and although they had heard scary rumors, they had no accurate image of it. Besides, their boat, *The Emerald Mile,* belonged to Kenton.

At this moment, Kenton told me, scores of passengers from motor rigs were walking around the rapid. During the previous two days, several big motor rigs had flipped in Crystal or had been trapped in the hole and stripped, scattering ninety-one passengers to swim, sometimes for miles, in the hellish section of river below. The Park Service had enacted a ruling that required guides to walk all passengers around Crystal. A ripple went through this crowd when one passenger spotted the (to them) silly little boat about to enter the biggest monster rapid in the known universe without even stopping to take a look at it. Niagara Falls would have made equal sense.

Kenton rowed down the tongue. He was worried about Crystal Hole but even more afraid of rowing too far right among the boulders to avoid it and staving a hole in the boat. This proved a misjudgment. The dory hit the first big diagonal, surfed left, then flipped in the hole, which now was big enough to gobble a pair of school buses. Farther down the rapid, in the submerged island, the upside-down dory collided with a rock and lost the upper foot of its stern and bow posts. Then, still upside down, all three boatmen and their wounded craft drifted out of view.

They righted the *Emerald Mile* above Tuna Rapid and rowed to shore to spend an hour patching it enough to continue. Then one of them started rowing again. Ultimately this tag team of three rowed the entire 237 miles of river (normally done in eleven to eighteen days) plus 40 miles of Lake Mead to the Grand Wash Cliffs in 36 hours and 38 minutes, a definite record (although Kenton still dreams of breaking it). This run broke their own previous record and also the one set by two brothers, Bob and Jim Rigg, who, on June 12, 1951, started rowing a Norm Nevills "sad iron" wooden boat from Lee's Ferry on nearly the same flow as above and arrived at Pierce Ferry 52 hours and 41 minutes later. Although the disparity between the two records seems great, Bob Rigg admitted to a group of us guides that Jim and he had spent about sixteen hours on shore to camp. Hence, their rowing times were identical.

Ironically, because the three had run the river illegally (due to Superintendent Marks's indecision), Marks decided to fine each of them $500 and suspend them from working in Grand Canyon National Park for two years. Their boss, Martin Litton, interceded on their behalf and persuaded Marks to drop the ban on working. Litton was not being altruistic. By then the river was flowing at 97,200 cfs. In fact, it rose to that level three days after we put in our regular trip and two days after the record-setting dory trip. A half dozen of the major rapids had grown gigantic beyond belief, and boats were flipping like crazy. Rudi, Kenton, and Steve were some of Litton's senior boatmen. He needed them to run his scheduled trips. And he got them.

I think each of them should have been awarded a golden oarlock instead of a slap in the face. Kenton wants to donate the *Emerald Mile* to the Smithsonian or the museum at Grand Canyon Village, but so far no bureaucratic decision-maker has expressed an interest in acquiring the rogue boat that set the undisputed speed record for traversing the Grand Canyon of the Colorado.

Unlike what the *Emerald Mile* faced, the last foreseeable challenge above Diamond Creek now was to claim a camp with early shade and protection from the storm now brewing in the west. Few beaches ahead fit both specifications.

The best was the upper beach at Mile 220. Most commercial trips taking out at Diamond Creek the next morning (instead of continuing down onto Lake Mead to take out at Pierce Ferry) aimed for this camp. Competition was certain. So, after Mile 217 Rapid, I rowed ahead of the other four boats as a hedge against the chance that a motor rig would blast past us at the last minute and pluck Upper 220-Mile from under our noses.

Behind me Fabry pulled a half-gallon plastic bottle of Canadian whiskey from a side box and gulped a swig for demonstration purposes. After everyone else on his boat had dosed themselves with this antiparasitical elixir, he snapped up the bottle, took another gulp, and struck the pose of a frozen quarterback about to nail a receiver entering the end zone. Then he sent the bottle spinning toward T.A.'s boat. They fished it out of the river and sampled it.

As I rowed ahead, the game of river whiskey football between the four boats behind me grew more animated—and sloppy. It would be a very happy beach when they got there. If they got there.

Passing Trail Canyon (Mile 219¼) I glanced back upstream. Our boats had dropped at least half a mile behind. Among them was a giant. I stared at it. The giant was pushing a bow wave. The giant had not stopped to socialize with T.A., the most sociable river guide on the Colorado. The giant wanted the camp at Upper 220.

I pivoted to aim my stern into Trail Rapid and row downstream. The beach at Upper 220 was now visible. And empty. But it was also nearly half a mile away, and half that distance was eddy water milling around like undecided molasses.

It seemed silly to row this hard for a camp, but I was trip leader and I wanted the best for our folks. And though the odds were small that any motor guide would crank his motor to top end for a mile just to yank that camp out from under our noses—most were too gentlemanly to pull such a stunt—the odds were not nonexistent. So I pulled on my oars. Hard.

Rowing through the molasses, I glanced at the approaching motor rig. Now we could not only watch the giant swell in size, we could hear his motor whining at redline. It was a race. John Henry versus the newfangled steam drill. I did not want to win and die, but I did want to win.

Thirty horsepower is a definite advantage, and my 50 percent lead was not quite enough handicap. I rowed anyway. Maybe it was. Once we broke into downstream current again, I thought I had a chance. The beach was so close now that the creosote bushes and mesquite trees looked like creosote bushes and mesquite trees. And the beach itself looked good. It had our names on it.

The racing motor rig plowed into Trail Rapid at full speed. But even now we could not see its pilot, just a dozen or so placid-looking passengers sitting immobile and staring ahead at the beach. The enemy.

It looked like we might make it—until the giant reached the foot of the eddy water and accelerated. But for us to lose, he would need to divert into the eddy to the right when we were within about eighty yards of the beach to pass us. Skunking us then is equivalent to pulling forward into a parallel parking spot after the driver ahead has stopped with his turn signal on and has begun backing into the same space. Rude.

But that's exactly what this pilot did. When he caught up, I waved and signaled toward the beach and then to us. The meaning was clear. Clear enough that he shook his head to say no, and without letting off a quarter inch on his throttle so we could speak verbally, pointed to the camp and pointed to himself, then blasted past us to the beach. Rude, but it was not the end of the world.

Annoyed, I rowed two hundred yards beyond this camp to the much more open Middle Camp, where the sun sets two hours later. In compensation, it offers an excellent view downstream four miles to the pyramidal silhouette of Diamond Peak. One glimpse of this mountain suddenly forced me to reassess my conclusion about the end of the world. The sky said the end might be imminent.

The western sky was charcoal, almost black. Lightning flickered like the electric tongue of some gigantic serpent. Thunder cracked like a public-service announcement to head for the nearest cave.

By the time I had secured our 150-foot safety line to a sturdy mesquite up the beach, the football team dribbled in to shore. I might have lost my race, but they had won their game. They tumbled from the boats in high spirits, half to find good camp spots and half to collect their black bags and tents. Smiles suddenly vanished as lightning cracked in the black sky. Everyone stopped and stared at Diamond Peak, now a dead ringer for Skull Island on the night when Fay Wray found herself married to a husband forty feet too big to handle.

As Rowan arranged their gear and started to set up their tent, Sheri (who had almost "died" on my boat during Lava Falls and who had watched the strange accident at Golf Ball Beach) said to me, "That looks like a serious storm."

"Maybe," I hedged, even though drops of rain were now pattering around us on the sand. "It's hard to predict whether it will keep moving in this direction. If we're lucky, it'll stay west of the Diamond Creek drainage."

"All that lightning . . . " she said, appealing for reassurance. "What are the chances that we could get hit by lightning?"

"About the same as getting hit in the head by a golf ball."

She jabbed me in the ribs.

But other possibilities were more likely than being struck by lightning—or a golf ball. If this storm was severe enough in the drainage of Diamond Creek, it could stop us from ending our trip there. A flash flood might mire the canyon and force us to continue downstream to Lake Mead across forty miles of reservoir to Pierce Ferry. And the storm could do even worse things than that.

Three weeks into July 1984, I was on a trip when the monsoons exploded with a vengeance. We had just hiked the sinuous, narrow cathedral of Blacktail Canyon at the head of the Aisle of the Conquistadores and were lounging in the shade under the ledges at its mouth. The weather had been threatening, but we had felt only a sprinkle.

Suddenly I heard Blacktail Rapid change its sound. Often the sound of a rapid changes as the water level fluctuates or the wind shifts, so I didn't think much about it. But the sound changed more in my ear aimed up Blacktail Canyon than in the ear aimed at the rapid. Just an echo, I figured.

But the roar grew louder. And louder. I looked around. None of the dozen or so people lounging here with me seemed to notice it. Even my oldest son, Mike, who was fourteen at the time, seemed oblivious. I suddenly knew what that roar was. It was not Blacktail Rapid. It was Blacktail Canyon.

"Flash flood!" I yelled. "Climb up these ledges *now!* Move it!" People scrambled as if Blacktail had filled wall to wall with rattlesnakes.

Instead of serpents, a four-foot wall of water rumbled around the bend, sending puffs of dust up before it as it spit cobbles and tumbled ahead of itself. The red slurry filled the canyon wall-to-wall, then quickly rose to about six feet up the walls. Surprisingly, it moved so slowly that a healthy person could have sprinted ahead of it. For maybe half a minute.

We watched it roar past in what had been a dust-dry bed and marveled at its force. I was impressed most with what an isolated phenomenon the heavy storm must have been that created this flash flood. All of us who have been here long enough have seen flash floods—huge spouts of red slurry roaring into muddy spray from the Redwall cliffs, dead bighorn sheep floating among the logs and flotsam and foam in midriver, rocks cracking loose from the cliffs to roar like cannon shots and kerplunk into the river. But even so, I never would have guessed that less than twenty-four hours earlier a much larger flash flood had gone roaring down Diamond Creek. Unfortunately, neither had our people at the mouth of Diamond.

Three days farther downriver, our trip was lounging (yes, again) in the cool shade of the Muav Amphitheater in Matka-tamiba Canyon (Mile 148), Dana Morris whirred his Park Service helicopter in low above the river between close walls and hovered above our boats, then a ranger tossed down a small plastic baggie half full of sand. Inside the baggie was a small typed note reading, "Diamond Creek flash flooded. Take out at Pierce Ferry."

Of course, we stopped at Diamond anyway to satisfy our curiosity. Diamond Creek flashed during most years, sometimes several times, but rarely seriously, and the Hualapai generally graded the road again within a day or two. It must have been a killer of a flash flood if they knew that five days of grading would not make it passable.

Yes, it must have been. Projecting at least fifty yards into the Colorado River from the mouth of Diamond Creek was a new delta of mud and boulders. The river flow had dropped last night about 4,000 cfs to reveal it.

Everyone jumped off the boats to sightsee. Michael Boyle led all of them up the canyon to see if the Hualapai had already graded a new road so that we would not have to row on down into Lake Mead. Alone, I walked out onto the delta.

I spotted green nylon buried in the boulders and mud. I dug it out. It was a fly for a Eureka tent—exactly the type we used, and still in good shape. A lucky find; now we had an extra fly. I walked farther onto the delta.

I spotted a military rocket box lid half buried in the sediment and rock. I wiggled it back and forth to yank it out. "COAL 1" it read, to signify that it now held charcoal briquettes. Connie, the wife of our area manager, Mike, had painted those words on it. This lid had definitely come off one of our boxes. How did they lose this rocket box off the truck?

Beyond it I found another lid, "EGG 2." It too was ours, or had been.

Feeling uneasy, I continued out toward the receding river. A cast-iron cooking griddle lay between two boulders. It was exactly the kind we used. It was squeaky clean and in perfect condition. Did they lose the whole load off the truck? I could not imagine how. No, that was ridiculous.

Then I spotted a twisted piece of red-painted steel, the same stuff that makes up the stake sides of our truck. I stopped and scanned around me. Here and there were other twisted pieces emerging from the mud. The answer seemed obvious, but I was hoping I was wrong.

My stomach sank as I walked up the creek. What the hell had happened to our truck, and to the crew?

The flash flood had really done a job on the creek. The road now existed only in our memories. I met Boyle and our group talking with two Hualapai men. Boyle spotted me and said excitedly, "You'll never guess what happened!"

"Our truck got hit by the flash flood and went down the drink."

He looked at me strangely. "How'd you know?"

272

"The delta," I explained. "Pieces of the truck are scattered all over it."

Later I learned that the only advance hint of this "event" had been intermittent thunderstorms the night before. During the day of their take-out, one week ago, it had drizzled for only a few minutes here at the river.

The road that day (July 20) was already a mess because in the previous twenty-four hours minor storms had carved gullies in it. River runners taking out were delayed because the road had been so hard to drive. Much of its final mile and a half to the river was slurrylike gravel.

Early that afternoon the passengers and half the crew of our other trip taking out here (Boyle's and my trip was staggered from theirs by a week) had ridden on the Hualapai bus up to Peach Springs. At 4:30 P.M. the remaining crew—Sam West, Greg Schill, Bill Brisbin, and Charles Rau—plus our area manager, Mike Walker, finally got the six boats and all the rest of the equipment loaded on our new two-ton GMC truck. We had a new GMC because only a year earlier our previous new GMC truck had caught fire and half melted into slag. Unaware that the region surrounding Peach Springs was buried under black clouds that had been pounding rain for a couple of hours, Walker started driving up the road behind a Dodge Power Wagon carrying equipment for another company, Outdoors Unlimited.

A half-mile up the creek, the Power Wagon had bogged down in the slurry of the "road" in the creekbed in a section of canyon between steep walls of schist called The Narrows. Debbie Jordan, driving the Dodge, and another woman, Amy, and guides Doug Carson and Dennis Silva couldn't get it moving. Walker, West, Brisbin, Rau, and Schill stopped and jumped out of the GMC to help.

The six men were about to push the Power Wagon when they heard a roar. Walker said the roar brought to mind a passing jet, thunder, and the Caterpillar the Hualapai used to grade the road.

But the sound was not right for any of those. Simultaneously he looked up at the bend of the canyon two hundred yards ahead and saw fingers of red-brown water five to ten feet high round the curve, rebound off the schist, and rumble toward them. The fingers swelled to a wall of slurry that quickly doubled its height, filling the canyon from wall to wall in a fifty-yard front.

"Run!" several people yelled. As everyone scattered, Walker ran back to the GMC, pulled the key from the ignition, rolled up the windows, then sprinted after the men heading for the high ground to the east.

When Debbie saw Walker rolling up the GMC's windows, she told Amy, "Don't get out. Roll up all the windows." Amy glanced again at the wall of water, then said, "No way," and groped for the door handle. This spurred Debbie to reconsider. Both girls leapt from the Power Wagon. Now the wall of water was a lot closer. Clad only in panties and T-shirts, they sprinted for the nearest cliff, to the west, and climbed up the weathered schist like squirrels until they reached an overhanging section of slick rock beyond which they could not climb. They had trapped themselves, but turning back now would have been suicide.

Diamond Creek, now a river flowing at least 5,000 cfs (instead of its usual 1 or 2 cfs) walloped both vehicles and engulfed them. Walker told me he knew the flash flood would hit both trucks hard, but he was hoping the GMC would stay where it was as the flood washed around it. But the wall of water picked up the Power Wagon like a leaf and tumbled it into the GMC. Within seconds both vehicles were swallowed and rolling downstream like toys. The trucks and all the equipment vanished almost instantly in the red slurry.

Guides from Arizona Raft Adventures (AzRA) were pulling their boats out of the Colorado upstream from the mouth of Diamond when they heard the roar. They stared at the sudden flood and for a brief instant spotted the dual wheels of the GMC surface in the flood as it rolled end-over-end in a twenty-five-foot wave into the Colorado. They were horrified. They were sure that our crew was still inside that truck.

Walker and the other men watched the flood roar past from a comfortable perch. The terrain allowed them to walk parallel to Diamond Creek in either direction. But Debbie and Amy, on the west side, could not have trapped themselves in a more frightening predicament. The overhanging slick rock stopped them from ascending higher or traversing laterally. The flash flood pounded past only a few feet below them, and, almost malevolently, it yawned directly below them in a gaping hole that appeared to be a duplicate of the one at Crystal. Not only were they trapped, they were trapped on an awkward cliff face that they must cling to or die. The flood would not subside for hours. If either of them lost her grip for even a second during those hours, she would die.

Unable to help them, Walker and the others hiked up to a ridge to view the entire flow as it rushed into the Narrows. In his haste to escape the flood, Schill had run right out of his flip-flops and now limped tenderly with his bare feet on the sharp schist and gneiss. (Soon the others went ahead of Schill, then sent a pair of flops back to him on loan via a runner.) Upstream the men saw the flood fill even the wider part of the canyon from wall to wall. They marveled at the unbelievable volume and power of this creek, which one could normally step over in a single stride. The pounding rain had transformed it into a full-sized river carrying a load of sediment that might fill a reservoir.

Walker watched the torrent for a while, then glanced at the red flood where the new GMC had vanished. He reached into the pocket of his Patagonia baggies and found the keys, pulled them out, looked at them, and said, "I guess we won't be needing these anymore." Then he tossed them into the racing slurry.

About an hour after the first wave, the AzRA guides hiked upstream with ropes, wondering if they would find anyone to rescue. The six men had settled into their ringside seat and were still trying to guess the volume of flow of the monster creek. Debbie and Amy might as well have been clinging to some crater wall on the moon.

The flood ebbed near dusk. An hour or so later, Sam West strapped on a couple of life jackets and swam across. On the other side he caught a rope, fixed it on the west side and set up a Tyrolean traverse to get the girls back to the east side. After their several hours of clinging for dear life to two-billion-year-old rock, West was Mr. Right. He rigged each of them into the traverse, and they were tugged across the subsiding flood to safety.

The AzRA crew hosted the get-together that night. Early the next morning almost everyone hiked up Diamond about five miles, where they met "rescue" vehicles. (Several months later, during winter and at very low water, a Hualapai spotted our GMC on the opposite side of the Colorado River well down Diamond Creek Rapid. Walker had put new front tires on the GMC only the day before losing it. The tires were still good.)

"But the Indians say," Boyle continued, "that no one got hurt."

"Good thing," I admitted, suddenly feeling a lot better.

We piled back into our boats and rowed down Diamond Creek Rapid and beyond it another dozen miles to Lake Mead, picking up stray gear in the eddies along the way. After an all-day, forty-mile tow that we shared with five Wilderness World boats and crews, Walker met us at Pierce Ferry, where we de-rigged and loaded our gear into our old truck. Forty miles down the road the truck broke down. Walker pulled his hair. "Anybody want my job?" he asked rhetorically.

Around midnight we hitched a ride on the Wilderness World van, grateful for their hospitality. About ten miles along Route 66 the van caught on fire. It was a propane-powered van, and the blaze had erupted next to the tank. I grabbed Mike and my ammo box, and all of us scrambled from the van like rats from a sinking ship, then sprinted a hundred yards into the dark desert before even glancing back.

276

The fire eventually burned out with no explosion, but the van had died in the process. We persuaded Kelly and Renee (both attractive enough for a male motorist to have a hard time passing them by) to hitch a ride into Kingman. For insurance I leapt out of the dark when a motorist finally stopped, and, pulling my son with me, rode along with them in the swaying, speeding car to line up some sort of vehicle to rescue everyone else stranded with the van and, miles behind it, with the truck.

Well, maybe a few challenges did remain at this point in the trip, I admitted to myself as I stared from Middle 220 at the ink-black sky behind Diamond Peak. I hope that son of a gun doesn't flash.

It hadn't, at least so far, and the weather this morning looked good. Already we had every boat de-rigged, washed, rinsed, stacked, and drying on shore at Diamond. Our piles of gear were ordered for quick loading, and the sky was almost blue. At the mouth of the creek a shiny, new, white GMC 2-ton truck with red stake sides appeared. Walker was driving.

"How was your trip?" he asked, grinning while we shook hands. Obviously he had already heard about the golf ball and the helicopter.

"Well, it wasn't boring."

"I guess *not*," he said. Then he lowered his voice, "Say, what's this with Danny?"

"I'll fill you in later. It's a good story. Say, did you pass the Hualapai bus on the way down?"

"No."

"Never mind," I said. "Here it comes."

I hustled over to the bus before Suzie even parked it. Suzie, a Hualapai, often took time away from tribal headquarters to drive the twenty-five-mile shuttle between here and Peach Springs on Route 66. She was a good driver. She was my favorite driver. Good drivers are nothing to take for granted.

Close cousins of the Havasupai, the thousand or so Hualapai (People of the Forest) once roamed over about 5 million acres between the Colorado and Bill Williams rivers. They subsisted on complex hunting and gathering cycles and practiced irrigation farming in the Hopi pattern. Immediately following the Civil War, though, Anglo ranchers and gold miners made their life hell, precipitating a war between the U.S. Cavalry and them that lasted three years. Due to the excellent logistical and tactical strategies of the Hualapai leaders, particularly Chief Cherum of the Middle Mountain subtribe, their 250 warriors held their own against immensely superior forces for years. Finally, in 1869 the cavalry launched a scorched-earth campaign on Hualapai camps and crops that forced a surrender.

In 1874 the army rounded up 667 Hualapais and shipped them to a reservation far south of their tribal lands at La Paz. There they were subjected to an enervating climate, epidemics of new diseases, and starvation rations from the federal government. In 1875 they escaped and fled north to their tribal territory, but by then their best lands had been usurped by Anglo ranchers. To survive, the Indians were forced to work for them as hired hands. In 1883 the federal government granted them a reservation of 900,000 acres (comprising the most useless 18 percent of their original lands). The tribe continued to decline due to disease and malnourishment. Anthropologist Tom McGuire quotes Oliver Gates (superintendent in 1905) as describing their reservation as "730,880 [*sic*] acres of the most valueless land on earth for agricultural purposes."

Today the Hualapai attempt to run their own affairs through an elected tribal council—whose decisions must be approved by the Bureau of Indian Affairs (in the Department of the Interior). Their land has not improved, employment among the work force is only about 25 percent, and alcoholism is epidemic. But the tribe earns money from logging their forests, from revenues for river trips and take-outs, and from incidentals such as permits for hunting eight to ten bighorn sheep annually. Still, much of their money is taxed to pay fat wages to Anglo officials of the BIA, which they resent. Currently the tribe is planning, in conjunction with the Las Vegas casino complex Circus Circus, to build a hotel, casino, and pleasure-boat marina near Mile 267 on Lake Mead.

"Hi, Suzie. How's life?" I asked past the folding door of the bus.

"Fine," she smiled, exhibiting a row of white teeth, the kind that make dentists miss their BMW payments. "How are you?"

"Great. Take-out went like clockwork."

"Are your people ready?" she asked, pulling her pad of forms to record for our respective bureaucracies the number of people she would be driving up to Route 66 at Peach Springs. "How many do you have?"

"Fifteen, plus three crew. I'll round them up right now." I ran to Diamond Creek to dunk myself in the 85-degree water, then started gathering everyone into the bus.

Now the bus was loaded with smiling and disheveled refugees from the twentieth century. In an hour we would be back in the land of telephones, traffic, checkbooks, credit cards, newspapers, alarm clocks, jobs, mortgages, and stores. Inescapable. Even the Hualapai had become partners in this culture.

Walker had brought an ice chest down. It sat on a spare seat. As if dictated by religion, our seven women sipped diet sodas while the men gulped cold beers. Suzie gunned the bus across the clear-flowing Diamond Creek, splashing a crescent bow wave that spattered the windshield. Everyone cheered.

I turned from the front, near Suzie, and said to everyone, "I want to thank you all for your great help with the de-rigging. We all appreciate it."

Everyone smiled, lifting cans in salute. I glanced out the rear window and saw the Colorado River as it vanished from view. Suddenly the cheer seemed misplaced and my stomach sank. Why did this canyon mean so much to me?

I turned around. Suzie navigated us through the Narrows,

where GMC Number Two had met its end in a red slurry as large as the Colorado at low water. (Had Bessie Hyde, aka Liz, hiked out through here on December 1, 1928?)

"I just want to remind everyone," I forced myself to announce, "that when you hear 'high side,' don't spill your drinks."

More cheers and hoots.

"How's that little boy of yours?" Suzie asked when the cheers subsided.

Before our daughter, Crystal, had been born—last fall—my wife, Connie, had often driven GMC Number Three down to meet us. And while she had helped get lunch ready for the folks, Suzie usually had bounced and played with little Cliff Hance. Although only a year old, he was already a social butterfly, and he liked women, especially Suzie. And with his red hair and blue eyes, Cliff must have seemed almost unreal to Suzie, whose own dark-eyed, black-haired children were uniformly what proper children should be. Despite this, or maybe because of it, I think Suzie was fascinated with Cliff.

"Great. I think," I admitted. Then it hit me. Cliff Hance Ghiglieri was about to have his second birthday. And I would be there.

I glanced again out the rear window of the bus. The Colorado River was hidden for good. I would be there, in the outside world, to celebrate Cliff's birthday, to be a father, a husband, a writer, a professor. But I would return here, and someday I would bring Connie and Cliff and Crystal with me.

Epilogue

I wrote this book to share what I consider to be one of the most impressive and fantastic experiences on Earth. I hope I have succeeded in communicating the wonder and reverence I feel for Grand Canyon well enough to spur you to experience it for yourself firsthand, if you have not done so already. Either way, I have an even dearer wish: that you will help to protect Grand Canyon and its desert ecosystem from needless destruction. If you would like to do so but do not know how, a good start would be to contact the following two organizations: Grand Canyon River Guides, P.O. Box 1394, Flagstaff, AZ 86002; and the Grand Canyon Trust, Route 4, Box 718, Flagstaff, AZ 86001. Next, write your member of Congress and the president. And don't forget to close your ammo box before you run that next rapid.

Selected References

Abbey, E. 1968. *Desert Solitaire*. New York: McGraw-Hill.

———. 1976. *The Monkey Wrench Gang*. New York: Avon.

Aitchison, S. 1985. *A Naturalist's Guide to Hiking the Grand Canyon*. Englewood Cliffs, N.J.: Prentice-Hall.

Akwesasne Notes. 1979. Native nations and the nuclear fuel cycle. Late winter, pp. 4–19.

Anderson, M. T. 1991. *North and South Bass Trails Historical Research Study, Grand Canyon National Park, Arizona*. Grand Canyon, Ariz.: U.S. Department of the Interior.

Arizona Daily Sun (Flagstaff). 1984. 20 escape flash flood in Grand Canyon. July 22, p. 30.

Awramik, A. M., and J. P. Vanyo. 1986. Heliotropism in modern stromatolites. *Science* 231:1279–1281.

Aydin, A., and J. M. DeGraff. 1988. Evolution of polygonal fracture patterns in lava flows. *Science* 239:471–476.

Baars, D. L. 1969. Major John Wesley Powell: Colorado River pioneer. In *Geology and Natural History of the Grand Canyon Region*, 10–18. Fifth Field Conference, Powell Centennial River Expedition 1969. Four Corners Geological Society.

Babbitt, B. 1978. *Grand Canyon: An Anthology*. Flagstaff, Ariz.: Northland Press.

Balsom, J. R. 1986. Application of heavy mineral analysis to Grand Canyon ceramics. *Western Anasazi Reports* 3 (4): 343–400.

Barnett, F. 1973. *Dictionary of Prehistoric Indian Artifacts of the American Southwest.* Flagstaff, Ariz.: Northland Press.

Bartimus, T. 1985. Colorado River: America's Nile flows uphill toward money. *Reno Gazette-Journal,* September 10, pp. 1C, 5C.

Beer, B. 1988. *We Swam the Canyon.* Seattle: The Mountaineers.

Bercovici, D., G. Schubert, and G. A. Glatzmaier. 1989. Three-dimensional spherical models of convection in the Earth's mantle. *Science* 244:950–955.

Billingsley, G. H. 1974. Mining in Grand Canyon. In W. J. Breed and E. C. Roat, eds., *Geology of the Grand Canyon,* 170–178. Flagstaff, Ariz.: Museum of Northern Arizona and the Grand Canyon Natural History Association.

Bishop, J., Jr. 1990. Glen Canyon Dam: Watching the watchers. *Arizona Republic* (Phoenix), March 25, p. C5.

Bolling, D. 1989. Water: North and south. *Headwaters* 8 (5): 2.

Bradley, G. Y. 1947. George Y. Bradley's Journal, May 24–August 30, 1869. Edited by W. C. Darrah. *Utah Historical Quarterly* 15: 31–72.

Breed, W. J., and E. C. Roat, eds. 1974. *Geology of the Grand Canyon.* Flagstaff, Ariz.: Museum of Northern Arizona and the Grand Canyon Natural History Association.

Brooks, J. 1950. *The Mountain Meadows Massacre.* 2d ed. Norman: University of Oklahoma Press.

Buechner, H. K. 1960. *The Bighorn Sheep in the United States: Its Past, Present and Future.* Wildlife Monographs No. 4. Washington, D.C.

Calloway, D. 1981. Neoplasms Among Navaho Children. Report submitted to the Division of Health Improvement Services, Navaho Tribe, Fort Defiance, Ariz.

Carr, M. H., R. S. Saunders, R. G. Strom, and D. E. Wilhelms, eds. 1984. *The Geology of the Terrestrial Planets (NASA SP-469).* Washington, D.C.: National Aeronautics and Space Administration.

Citizens for Environmental Responsiblity. 1988. *The Grand Canyon Is Under Siege: Uranium in Coconino County; A Fact Sheet.* Flagstaff, Ariz.: Citizens for Environmental Responsibility.

Colorado River Studies Office. 1990. What is the Glen Canyon Dam EIS all about? *Colorado River Studies Office Newsletter* 1:4.

Cooley, J. 1988. *The Great Unknown: The Journals of the Historic First Expedition Down the Colorado River.* Flagstaff, Ariz.: Northland Press.

Council on Environmental Quality and the Environmental Protection Agency. 1989. *Facts About the National Environmental Policy Act.* Washington, D.C.: United States National Environmental Protection Agency.

Crampton, C. G. 1986. *Ghosts of Glen Canyon.* St. George, Utah: Publishers Place.

Crawford, M. 1990. Scientists battle over Grand Canyon pollution. *Science* 247:911–912.

Crumbo, K. 1981. *A River Runner's Guide to the History of the Grand Canyon.* Boulder, Colo.: Johnson Books.

Dagget, D. 1987. Uranium Rush. *Amicus Journal* 9 (3): 5–6.

D'Azevedo, W. L., ed. 1986. *Handbook of North American Indians.* Vol. 11: *Great Basin.* Washington, D.C.: Smithsonian Institution.

Dellenbaugh, F. S. 1904. *The Romance of the Colorado River.* Chicago: Rio Grande Press.

———. 1908. *A Canyon Voyage: The Narrative of the Second Powell Expedition.* Reprint. 1984. Tucson: University of Arizona Press.

DeVoto, B., ed. 1953. *The Journals of Lewis and Clark.* Boston: Houghton Mifflin.

Dillehay, T. D. 1990. *Monte Verde: A Late Pleistocene Settlement in Chile.* Vol. 1: *Paleoenvironment and Site Context.* Washington, D.C.: Smithsonian Institution.

Dominy, F. E. 1965. *Lake Powell: Jewel of the Colorado.* Washington, D.C.: U.S. Government Printing Office.

Environmental Defense Fund. 1990. In Grand Canyon air pollution case Interior Department is both victim and villain. *EDF Letter* 21 (1): 6.

———. 1990. *Estimates of Economic Impacts of Implementing Interim Flow Release Patterns at Glen Canyon Dam.* Environmental Defense Fund report for the U. S. House of Representatives Subcommittee on

Water, Power and Offshore Energy Resources. July 12, 1990. Oakland, Calif.

Fiedel, S. J. 1987. *Prehistory of the Americas*. New York: Cambridge University Press.

Fradkin, P. L. 1981. *A River No More*. New York: Knopf. Reprint. 1984. Tucson: University of Arizona Press.

Fritts, H. C. 1972. Tree rings and climate. *Scientific American* 226 (5): 92–100.

Garrett, W. E. 1978. Grand Canyon: Are we loving it to death? *National Geographic* 154 (1): 2–51.

Ghiglieri, M. P. 1988. *East of the Mountains of the Moon*. New York: Macmillan, Free Press.

————. 1992. Screenplay for *River of Stone: The Powell Expedition*. Salt Lake City: Utah Public Television KUED-7. Film.

Goetz, A.F.H., F. C. Billingsly, A. R. Gillespie, M. J. Abrams, R. L. Squires, E. M. Shoemaker, I. Lucchitta, and E. R. Elston. 1975. *Application of ERTS Images and Image Processing to Regional Geologic Problems and Geologic Mapping in Northern Arizona*. NASA Technical Report 32-1597. Pasadena, Calif.: California Institute of Technology, Jet Propulsion Laboratory.

Goldstein, S. 1989. Interior Secretary Lujan directs Bureau of Reclamation to prepare an environmental impact statement of Glen Canyon Dam operations. U. S. Department of the Interior news release.

Groves, D. I., J.S.R. Dunlop, and R. Buick. 1981. An early habitat of life. *Scientific American* 245 (4): 64–73.

Guse, N. G. 1974. Colorado River bighorn sheep survey. *Plateau* 46 (4): 135–138.

Hall, A. 1948–49. Three Letters by Andrew Hall. Edited by W. C. Darrah. *Utah Historical Quarterly* 16–17:505–508.

Hall-Quest, O. 1969. *Conquistadors and Pueblos: The Story of the American Southwest, 1540–1848*. New York: E. P. Dutton.

Hamblin, W. K., and J. K. Rigby. 1968–69. *Guidebook to the Colorado River*. 2 vols. Provo, Utah: Brigham Young University.

Harker, V. 1988. Ruling may imperil sacred Indian lands. *Arizona Republic* (Phoenix), April 24, pp. B1, B3.

Heppenheimer, T. A. 1987. Journey to the center of the earth. *Discover* 8 (11): 86–93.

Hevly, R. H. 1988. Paleoenvironments and historic occupation of the Colorado Plateau. Manuscript, Northern Arizona University, Department of Biological Sciences.

Hill, P. 1982. Letter to the editor. *Northland News,* June 23.

Hille, J. 1982. Jamaican connection: Spirit of reggae thrives at the bottom of the Grand Canyon. *Arizona Republic* (Phoenix), October 31, pp. D1, D7.

Hoffman, P. F. 1989. Speculations on Laurentia's first gigayear (2.0–1.0 Ga). *Geology* 17:135.

Hoffmeister, D. F. 1971. *Mammals of Grand Canyon.* Chicago: University of Illinois.

Howe, J. 1986. *Troubled Waters.* Salt Lake City: Utah Public Television KUED-7. Film.

Howland, O. G. 1947. Letters of O. G. Howland to the *Rocky Mountain News. Utah Historical Quarterly* 47:95–105.

Hughes, J. D. 1978. *In the House of Stone and Light.* Grand Canyon Natural History Association.

Jones, A. T., and R. C. Euler. 1979. *A Sketch of Grand Canyon Prehistory.* Grand Canyon Natural History Association.

Kabotie, F., and B. Belknap. 1977. *Fred Kabotie: Hopi Indian Artist.* Flagstaff, Ariz.: Museum of Northern Arizona and Northland Press.

Kelly, K. W., ed. 1988. *The Home Planet.* New York: Addison-Wesley.

Kolb, E. L. 1914. *Through the Grand Canyon from Wyoming to Mexico.* New York: Macmillan. Reprint. 1990. Tucson: University of Arizona Press.

Lavender, D. 1985. *River Runners of the Grand Canyon.* Grand Canyon: Grand Canyon Natural History Association.

Lekson, S. H., T. C. Windes, J. R. Stein, and W. J. Judge. 1988. The Chaco Canyon community. *Scientific American* 259 (1): 100–109.

Lucchitta, I. 1984. Development of landscape in northwestern Arizona: The country of plateaus and canyons. In T. L. Smiley, J. D. Nations, T. L. Pe'we', and T. L. Schafer, eds., *Landscapes of Arizona: The Geological Story,* 269–301. New York: University Press of America.

————. 1988. Canyon maker: A geological history of the Colorado River. *Plateau* 59 (2): 2–32.

McGuire, T. R. 1983. Walapai. In A. Ortiz, ed., *Handbook of North American Indians,* vol. 10: *Southwest,* 25–37. Washington, D.C.: Smithsonian Institution.

McKee, E. D., R. F. Wilson, W. J. Breed, and C. S. Breed. 1967. *Evolution of the Colorado River in Arizona.* Museum of Northern Arizona Bulletin 44. Flagstaff: Museum of Northern Arizona.

McPhee, J. 1971. *Encounters with the Archdruid.* New York: Farrar, Straus and Giroux.

Mahoney, J., and P. Hogan. 1988. Politics on sacred land. Manuscript in possession of the author.

Martin, P. S., and R. G. Klein, eds. 1984. *Quaternary Extinctions: A Prehistoric Revolution.* Tucson: University of Arizona Press.

Martin, R. 1989. *A Story that Stands Like a Dam.* New York: Henry Holt.

Matlock, G. 1988. *Enemy Ancestors.* Flagstaff, Ariz.: Northland Press.

Meyer, R. E. 1983. Raging Colorado flood batters Grand Canyon: Beaches destroyed, wildlife killed as melting snow sends river over old high-water marks. *Los Angeles Times,* June 25, pp. 1, 14, 16, 17.

Moley, R. 1955. *What Price Federal Reclamation.* New York: American Enterprise Association.

Monson, G., and L. Sumner. 1980. *The Desert Bighorn.* Tucson: Univ. of Arizona Press.

Morrison, R. 1988. *Memorandum: Sediment in the Upper Colorado River Basin and Lake Powell.* Salt Lake City, Utah: U. S. Department of the Interior, Bureau of Reclamation, Upper Colorado Region.

Nance, R. D., T. R. Worsley, and J. B. Moody. 1988. The supercontinent cycle. *Scientific American* 259 (1): 72–79.

National Academy of Sciences. 1987. *River and Dam Management: A*

Review of the Bureau of Reclamation's Glen Canyon Environmental Studies. Washington, D.C.: National Academy Press.

National Broadcasting Company. 1987. "Unsolved Mysteries," episode entitled "Mysterious Legends: 1928, the Grand Canyon." November 29. Film.

Nelson, L. 1990. *Ice Age Mammals of the Colorado Plateau*. Flagstaff: Northern Arizona University.

Noble, D. G., ed. 1987. *Exploration: Wupatki and Walnut Canyon*. Santa Fe: School of American Research.

Oppelt, N. T. 1981. *Guide to Prehistoric Ruins of the Southwest*. Boulder, Colo.: Pruett.

Ortiz, A., ed. 1979. *Handbook of North American Indians*. Vol. 9: *Southwest*. Washington, D.C.: Smithsonian Institution.

———. 1983. *Handbook of North American Indians*. Vol. 10: *Southwest*. Washington, D.C.: Smithsonian Institution.

Plog, F. 1979. Prehistory: Western Anasazi. In A. Ortiz, ed., *Handbook of North American Indians*, vol. 9: *Southwest*, 108–130. Washington, D.C.: Smithsonian Institution.

Porter, E. 1963. *The Place No One Knew*. San Francisco: Sierra Club Books.

Powell, J. W. 1879. *Report on the Lands of the Arid Region of the United States, with a More Detailed Account of the Lands of Utah*. Reprint. 1983. Boston, Mass.: Harvard Common Press.

———. 1895. *The Exploration of the Colorado River and Its Canyons*. Reprint. 1961. New York: Dover.

———. 1947. Major Powell's Journal, July 2–August 28, 1869. Edited by W. C. Darrah. *Utah Historical Quarterly* 15:125–131.

Powell, S., and G. J. Gummerman. 1987. *People of the Mesa: The Archeology of Black Mesa, Arizona*. Tucson, Ariz.: Southwest Parks and Monuments Association; and Carbondale: Southern Illinois University Press.

Prescott, W. H., J. L. Davis, and J. L. Svarc. 1989. Global positioning system measurements for crustal deformation. *Science* 244:1337–1340.

Rahm, D. A. 1974. *Reading the Rocks*. San Francisco: Sierra Club Books.

Rasmussen, D. I. 1941. Biotic communities of the Kaibab Plateau. *Ecological Monographs* 11 (3): 229–275.

Reed, M. 1989. Our uranium legacy. *Northern Arizona Environmental Newsletter* 1 (1): 3–6.

Reisner, M. 1986. *Cadillac Desert: The American West and Its Disappearing Water*. New York: Viking.

Ritter, D. F. 1967. Rates of denudation. *Journal of Geological Education*. 15 (4): 154–159.

Rosewicz, B. 1989. Grand Canyon air pollution tied to facility: EPA blames power plant partly owned by U. S. for poor winter views. *Wall Street Journal*, August 30, p. A4.

Rusho, W. L. 1981. *Desert River Crossing*. Salt Lake City: Peregrine Smith.

———. No date. *Flood Waters of the Colorado River System: Damage and Repair of the Spillways at Glen Canyon Dam*. Salt Lake City: U. S. Department of the Interior, Bureau of Reclamation, Upper Colorado Region. Film.

Russell, D. 1987. The Monkeywrenchers. *Amicus Journal* 9 (4): 28–42.

Schwartz, D. W. 1969. Grand Canyon prehistory. In *Geology and Natural History of the Grand Canyon Region*. Fifth Field Conference, Powell Centennial River Expedition, 1969, pp. 35–40. Four Corners Geological Society.

———. 1983. Havasupai. In A. Ortiz, ed., *Handbook of North American Indians*, vol. 9: *Southwest*, 13–24. Washington, D.C.: Smithsonian Institution.

Schwartz, D. W., R. C. Chapman, and J. Kemp. 1980. *Archaeology of the Grand Canyon: Unkar Delta*. Grand Canyon Archaeology Series, vol. 2. Santa Fe: School of American Research Press.

Simmons, M. 1979. History of Pueblo-Spanish relations to 1821. In A. Ortiz, ed., *Handbook of North American Indians*, vol. 9: *Southwest*, 178–193. Washington, D.C.: Smithsonian Institution.

Skow, J. 1988. The Forest Service follies. *Sports Illustrated* 68 (11): 76–88.

Spier, L. 1928. Havasupai ethnography. *Anthropological Papers of the American Museum of Natural History* 29 (3): 81–392.

Stanton, K. 1988. Invading the last frontier: Arizona's north country, home to booming uranium mines and cultural plundering of Indian graves, gets a bitter taste of "progress." *New Times,* February 17–23, pp. 22–35.

Stanton, R. B. 1932. *Colorado River Controversies.* New York: Dodd, Mead and Company.

———. 1965. *Down the Colorado.* Norman: University of Oklahoma Press.

Stegner, W. 1954. *Beyond the Hundredth Meridian.* Cambridge: Cambridge University Press, Riverside Press.

Stevens, J. E. 1988. *Hoover Dam: An American Adventure.* Norman: University of Oklahoma Press.

Stevens, L. 1983. *The Colorado River in Grand Canyon: A Guide.* Flagstaff, Ariz.: Red Lake Books.

Sumner, J. C. 1947. J. C. Sumner's Journal, July 6–August 31, 1869. Edited by W. C. Darrah. *Utah Historical Quarterly* 15:113–123.

Tessman, N. 1986. "I wonder if I shall ever wear pretty shoes again": The disappearance of Glen and Bessie Hyde. *Sharlot Hall Gazette* (Prescott, Ariz.) 13 (1): 1–6.

Thybony, S. 1985. What happened to Bessie Hyde? *Outside* 5 (10): 108.

———. 1987. *Fire and Stone: A Road Guide to Wupatki and Sunset Crater.* Tucson, Ariz.: Southwest Parks and Monuments Association.

———. 1988. *Walnut Canyon National Monument.* Tucson, Ariz.: Southwest Parks and Monuments Association.

———. 1989. A river mystery. In *First Descents: In Search of Wild Rivers,* ed. C. O'Connor and J. Lazenby. Birmingham, Ala.: Menasha Ridge Press.

Trepper, B. 1984. Skanking with the Havasupai. *Reggae & African Beat* 3 (2): 12–15, 45.

Turner, C. G., II. 1983. Taphonomic reconstruction of human violence and cannibalism based on mass burial in the American Southwest. In G. M. LeMoine and A. S. MacEachern, eds., *Carnivores, Human Scavengers and Predators,* 219–240. Calgary: Archaeological Association of the University of Calgary.

————. 1989. Teec Nos Pos: More possible cannibalism in northeastern Arizona. *Kiva* 54 (2): 147–152.

Turner, C. G., II, and J. A. Turner. In progress. Postmortem damage to PIII human skeletal remains from the Wupatki mass burial (NA405, room 59), and the House of Tragedy (NA682), Wupaki National Monument, Arizona.

Udall, S. L. 1988. A letter from Mineral Policy Center Board chairman. *Clementine* (Autumn): 1.

————. 1990. The trails of Juan de Oñate and Francisco Garcés. *Arizona Highways* 66 (7): 4–15.

U.S. Department of Agriculture. National Forest Service. Case No. 1874 (February 25, 1987). *RE: Canyon Mine—Appeal of the Havasupai Tribe*. Before the United States Department of Agriculture, National Forest Service, Office of the Chief, Washington, D.C.

U.S. Department of the Interior. 1988. *Glen Canyon Environmental Studies: Final Report, January 1988*. Salt Lake City: U. S. Department of the Interior.

————. 1988. *Glen Canyon Environmental Studies: Executive Review Committee Final Report, May 1988*. Salt Lake City: U. S. Department of the Interior.

U.S. Department of the Interior. Bureau of Reclamation. 1990. *Glen Canyon Dam Environmental Impact Statement: Development of Alternatives, March 1990*. Salt Lake City: U. S. Department of the Interior, Bureau of Reclamation.

U.S. Department of the Interior. Bureau of Reclamation. Durango Projects Office. 1987. *Lake Powell 1986 Sediment Study Report*. Durango, Colo.: Bureau of Reclamation, Durango Projects Office.

Walters, J. E. 1979. Bighorn sheep population estimate for the South Tonto Plateau—Grand Canyon. Manuscript. National Park Service, Grand Canyon National Park. 12 pp.

Walters, J. E., and R. M. Hansen. 1978. Evidence of feral burro competition with desert bighorn sheep in Grand Canyon National Park. *Desert Bighorn Council Transactions* 22:10–16.

Waters, F. 1963. *Book of the Hopi*. New York: Ballantine.

Wegner, D. 1990. Environmental Impact Statement. *Grand Canyon River Guides* 3 (1): 3.

Weise, W., ed. 1981. Birth defects in the Four Corners area. Transcript of proceedings of a conference sponsored by the University of New Mexico, Department of Family, Community, and Emergency Medicine, Albuquerque, February 27, 1981.

Whitney, S. 1982. *A Field Guide to the Grand Canyon*. New York: Morrow.

Wilcox, D. R. 1988. Hohokam warfare. Manuscript. Northern Arizona University, Department of Anthropology.

Worster, D. 1985. *Rivers of Empire: Water, Aridity, and the Growth of the American West*. New York: Pantheon.

Index

Grand Canyon along, 264–67; fate of, 78; first footage shot of rapids of, 47–48; gradient and length of, in Grand Canyon, 41; named by, 38; peak flows of, 6, 42; rapids in, 40, 41–42; shortage of water in (as political "deficit" river), 73, 76; silt load of, 30, 41, 42, 160; size of, 74; storage capacity of dams along, 75; temperature of, in Grand Canyon, 46, 188; trout fishery in, 25, 30; ultimate use of, 79; and James White, 242–43. *See also* Bureau of Reclamation; Crystal Rapid; Glen Canyon Dam; Grand Canyon; Hance Rapid; Lava Falls Rapid; Lava flows and volcanism

Colorado River Compact of 1922, 72–73, 181

Colorado River Storage Act, 83

Colorado River Storage Bill, 81

Congressional Mandate of August 25, 1916, 81

Continental drift: cause of, 66–68; and Colorado Plateau, 61, 62, 64–68; continental breakup due to, 67; convection as cause of, 67; plate tectonics of, 65–68; radioactive decay in, 66; red granites as evidence of, 68; seafloor spreading in, 65–66; and thickness of continental plates, 67; and thinness of oceanic plates, 66–67; volcanism as consequence of, 66. *See also* Global Positioning System; Laurentia

Coronado, Francisco Vásquez de: expedition of, 133–35

Cortez, Hernán, 134

Cottonwoods, 164, 166, 175, 176, 226

Crampton, C. Gregory, 183

Crossing of the Fathers, 25, 27

Cruise, Tom, 227–28

Crumbo, Kim, xvi, 10

Crystal Creek, 4, 162

Crystal Rapid: accidents in, 9–10, 12–14, 149, 265–66; air collisions over, 21; Crystal Hole of, 13, 17, 18, 19–20, 275; description of, 1, 10–11, 15; evolution of, 4–5, 9–10; flood stage at, 10–11; navigation of, 12, 16–20; New Wave in, 12, 16, 18; research at, 191; Rock Garden in, 13, 19, 20; scouting of, 11–12, 14–15; Slate Creek diagonal of, 13, 17, 18; Thank God Eddy of, 21

Custer, George Armstrong, 87

Dagget, Dan, xvi, 218, 219, 220, 224

Daggett, John, 243

Dale, Eban, 206–7

Dale, O'Connor, xvi, 203–4, 205–6

Dale, Regan, xvi, 205–6

Danny, 237, 239–40, 248, 249, 250–52, 260, 262, 277
Darwin, Charles, 65
Davis Dam, 75, 82
Deer Creek, 121, 170
Deer Creek Falls, 177, 178, 179
Deer Creek Grotto, 175–77
Deer Creek Rapid, 179
Deer Creek Spring, 170
Dendrochronology, 123–24, 129
de Niza, Fray Marcos, 133–34
Denver Colorado Canyon and Pacific Railroad, 157, 162
Department of Agriculture, 111
Desert Archaic culture, 106–7, 108, 109, 163
Diabase Sill, 139–40, 164
Diamond Creek, 41, 175, 203, 204, 205, 206, 208, 267; flash flood in, 270, 271–76; temperature of, 279; as tributary to Colorado system, 172
Diamond Creek Rapid, 276
Diamond Peak, 269, 277
Dinosaur National Monument, 80, 181
Domar riverboats, 16, 19, 33–34, 141, 254
Domínguez, Fray Francisco Atanasio, 26, 27
Dominy, Floyd, 77, 80, 81, 82, 83, 84
Douglas, Senator Paul, 75
Dox Sandstone, 112, 119, 121, 139; age of, 112; as building material, 122; erodability of, 119–20; size of, 119

Dunn, William, 87, 88; abandons Powell Expedition, 93, 96–101, 103–4; abused by Powell, 96–97, 98; fate of, 100–101, 103–4; thrown in river, 89, 90, 92

Earth First!, 84, 186
Earth Island Institute, 84
East Kaibab Monocline, 114
Echo Cliffs, 24, 35, 50
Echo Cliffs Monocline, 25, 26, 34
Echo Park, 181
Eddy, Clyde, 48
Ediacarian radiation, 113
Elk, 108, 217, 222
Emerald Mile, 265–67
Emma Dean, 88, 89, 95; abandonment of, 99; flips of, 89, 90, 92
Energy Fuels Nuclear, 219–21, 222–23, 224; number of uranium mines planned by, 220; uraniums spills by, 222
Environmental Defense Fund, ELFIN model, 193–94
Erosion, 59; of North American continent, 113; in Grand Canyon, 171–75, 234. *See also* Glen Canyon Dam
Escalante, Fray Velez de, 26, 27, 136
Esplanade member of Supai Group, 158, 170, 212, 235, 237, 247
Esteban, 133–34

Euler, Bob, 218

Fabry, Michael, 12–16, 31–33, 35, 85, 141, 215, 237, 239–40, 248–49, 251, 252, 261–62, 267
Fancher, 28
Federal Mining Law of 1872, 220, 224
First Granite Gorge, 1, 64; description of, 143–44; and Glen and Bessie Hyde, 203; and Powell Expedition, 91–92
Flagstaff, 29, 32, 129, 143, 144, 215, 218, 222, 224, 231, 243
Flagstaff Medical Center, 146, 255, 259
Flash floods, 270, 271–76
Folsom hunters, 107, 108
Ford, Gerald, 77
Forster Rapid, 154
Fossil Rapid, 154
Fossils, 49, 62, 64
Fradkin, Philip L., 81, 82, 181, 182–83
Friends of the Earth, 84
Friends of the River, 225
Furnace Flats, 118, 119, 120, 126

Galeros Formation, 114, 115, 116, 117, 118
Galloway, Nathaniel, 46
Garcés, Francisco Tomás, 38, 136, 217; death of, 217
Garrett, W. E., 187, 217
Gates, Oliver, 278

Gellenson, Art, 208
Geology: astrophysics of, 63; carving of Grand Canyon, 170–75; dating, 64; erosion in, 234–35; genesis of Paleozoic Sequence in Grand Canyon, 59–62; igneous rock in, 63; Iron Catastrophe of, 63; metamorphism in, 63–64; paleomagnetism of, 64, 65; radioactive decay in, 66; radiometric dating of, 64; sedimentation in, 61–62, 63; transgression and regression of seas in, 61, 67; unconformities in, 61. *See also* Continental drift, Pangaea
Ghiglieri, C. Michael, 140, 144, 147, 270, 276
Ghiglieri, Cliff, H., 53, 148–49, 197, 215, 249, 264, 280
Ghiglieri, Connie, 53, 144–46, 147–49, 215, 249, 264, 280
Ghiglieri, Crystal, 53, 197, 249, 264, 280
Ghiglieri, Michael P., career of, 52–53
Glen Canyon: beauty of, 80, 180, 184–85; capacity of Lake Powell in, 180; crossing points in, 27; gradient of Colorado River within, 180; lessons from, 181; effect of Marble Canyon Dam upon, 71; Powell Expedition in, 90; rapids in, 180; resources lost

from, 181, 183, 184; trout fishery in, 30

Glen Canyon Dam, 42, 75, 263; agreement to build, 80; bedrock of, 7, 182–83, 184; capacity to spill water of, 8; changes in Colorado River due to, 30, 34, 36, 40, 42, 46, 187–95; changes in Grand Canyon by, 40, 188–93; cost to build, 188; cost of environmental damage from, 187–88; cost of environmental research for, 192; and CREDA, 189, 193; death toll among construction workers at, 188; design of, 184; impacts on endangered fish by, 190; hydropower and, 188, 189, 191, 193–94; and National Environmental Policy Act (NEPA), 188; operations of, 5–7, 8–9, 30, 70, 191, 192; peaking power of, 30, 189; purpose of, 6, 182, 193; and Rainbow Bridge, 183; and rapids, 190; and recommendations by GCES, 194–95; and San Juan River, 184; seepage around, 184; siltation of, 187; size of, 182, 184; spillways of, 6–7, 9; and WAPA, 189, 193; and water loss, 187; effect on whitewater boating of, 190. *See also* Bureau of Reclamation; Glen Canyon; Grand Canyon Glen Canyon Environmental

Studies ((GCES), 191–95. *See also* Glen Canyon Dam

Global Positioning System, 66

Gneiss Canyon, 203

Goddard, Renee, 13, 14, 258, 277

Goldwater, Senator Barry, 77, 187

Golf, 248–49

Goodman, Al, 221

Goodman, Frank, 87, 88, 89

Grand Canyon: age of, 171, 174; air crashes over, 21, 117; and Congress, 83; natural dams in, 235; death toll in, 21; depth of, 44, 174–75, 236; discovery of, 134–35, 151; erosion in, 234–35; exploration of, 86, 90–104; extinct mammals in, 108; formation of, 170–75; heat in, 263; landslides in, 170; and marriages, 200; numbers of river runners in, 36; precipitation in, 4, 56, 120; romance of, 199; routes into, 212; spiritual effect of, 22–23; trout fishery in, 25, 30; visitors to, 83; volume of, 44

Grand Canyon Clinic, 145

Grand Canyon Game Reserve, 155

Grand Canyon Guides Association, 281

Grand Canyon National Park: air quality problems in, 83, 222; date established, 42; re-

Hawikuh, 133–34
Hawkins, William Rhodes, 87; at Lava Cliffs Rapid, 99; and Separation Canyon, 96–97, 99
Hayden, Senator Karl, 77, 81
Hayduke, George, 186
Helin, Bruce, 14, 19
Hell's Canyon (Idaho), 43
Helmick, Earl, 200, 208–9
Henderson, Royal Bart, 258–59
Hermit Rapid, 202
Hermit Shale, 59, 61, 62, 219
Hetch Hetchy Valley, 81
Hevly, Richard, 129
Hill, Pam, 220
Hoffman, Paul, 64
Hoover, Herbert, 72
Hoover Dam, 9, 41; cost of, 73; death toll of workers at, 74; as electrical generator, 74, 182; illegal attachment of revenues from, 82; siltation of, 187; size of, 73–74,
Hopi, 120, 216; agriculture of, 122, 125; ancestral name of, 132; destruction of Awatovi by, 136; Cha-Kwaina Katsina of, 134; Chief Chipiya of, 132; exodus from Grand Canyon by, 131; in Grand Canyon, 109–10; as guides of Cardenas, 134, 151; and Havasupai, 223; name of, 132; population of, 131, 136; pottery of, 132; in Pueblo Rebellion of 1680, 136; Rastafari-

anism among, 231; and Red Butte, 223; reservation of, 137; salt mines of, 109–10, 111; and San Francisco Peaks, 132; and Sinagua, 132; Sipapu of, 131, 218; at Tusayan, 134, 135; and Walnut Canyon, 132; warfare against Spanish by, 136; and Wupatki, 131–32. *See also* Anasazi, Cárdenas, Lt. García López de; Coronado, Francisco Vásquez de
Horn Creek Rapid, 5, 206; *Sweet Marie* destroyed in, 162
Hotauta Conglomerate, 139
House Rock Rapid, 90
Howland, Harry, 201
Howland, Oramel G., 87; at Disaster Falls, 88, 89; fate of, 100–101, 103–4; abandons Powell Expedition, 93, 96–101, 103–4
Howland, Seneca, 87; at Disaster Falls, 88; fate of, 100–101, 103–4; abandons Powell Expedition, 93, 96–101, 103–4
Hualapai, 35, 204, 217, 271, 272, 273; and Bureau of Indian Affairs, 278–79; Chief Cherum of, 278; and desert bighorn, 155, 278; employment among, 278; farming by, 278; meaning of name of, 277; reservation of, 278; original territory of, 277–78; war with U.S. Cavalry of, 278

North Kanab Uranium Mine;
Rio Puerco
Ute Ford. *See* Crossing of the
Fathers
Utes, 109

Vasey's Paradise, 102
Vaughn Spring, 175
Verde-Salt-Gila river system, 75
Vermilion Cliffs, 24, 35
Vermilion Cliffs Restaurant, 31
Viking I and II (NASA space
probes), 43
Violet-green swallows, 38
Vishnu Schist, 64, 68, 112, 143,
150; age of, 64, 112; meta-
morphosis of, 143–44; as
mountain range, 64–65
Volcanoes, 66. *See also* Conti-
nental drift; Lava flows and
volcanism in Grand Canyon
Vulcan's Anvil, 237

Waghi River (Papua New
Guinea), 13, 145
Walhalla Plateau, 115, 125
Walker, Connie, 272
Walker, Michael, xvi, 147, 272,
273–74, 275, 276, 277
Walnut Canyon, 129–30
Washington Post, 82
Watahomigie, Tobey, 229
Watt, James, 191

Watut River (Papua New
Guinea), 145
Wegener, Alfred, 65
Wegner, David, xvi, 191–93,
195
West, Sam, 273, 276
Western Area Power Adminis-
tration (WAPA) 30, 189, 190
Western whiptail lizard, 226
Whale (Curtis Hanson), 238,
239
White, Georgie, 9–10, 29
White, James, 242–43
Whitmore Wash, 263
Wilcox, David, 131
Wilderness World, 276
Wilford (Idaho), 7–8
Willoughby, Otis, 235–36
Wilson, Woodrow, 42
Worster, Donald, 75, 79, 80
Wray, Fay, 269
Wupatki, 128, 129–30, 131–32

Yampa River, 80; dams planned
on, 181
Yavapai, 217
Youghiogheny River, 245
Young, Brigham, 27, 28, 29
Yumans, 217
Yurok, 225

Zoroaster Granite, 20, 68, 143,
150
Zuni, 128, 133, 133–34

About the Author

MICHAEL P. GHIGLIERI holds a B.S. and an M.A. in biology and a Ph.D. in biological ecology from the University of California, Davis. In addition to conducting research on gorillas and chimpanzees in Uganda and serving as an assistant professor at Northern Arizona University, he has been a professional river guide for eighteen years, with experience on three dozen rivers worldwide. His interest in Grand Canyon dates from 1976 when he was rowing his first trip down the Colorado. Over the years he has become convinced that the Grand Canyon Colorado may well be the most impressive river trip on earth. His scientific writing includes articles in *Scientific American*, *Discover*, the *Journal of Human Evolution*, the *African Journal of Ecology*, and *Oryx*, and he has published articles in *Backpacker*, *River Runner*, *River World*, and other magazines. Besides *Canyon*, his books include *The Chimpanzees of Kibale Forest*, *East of the Mountains of the Moon*, and *The Dark Side*. Ghiglieri is currently planning a research project on the desert bighorn in Grand Canyon.